HOMEMADE CLEANERS

DEC   2013

HOMEMADE CLEANERS

QUICK-AND-EASY, TOXIC-FREE RECIPES
TO REPLACE YOUR KITCHEN CLEANER, BATHROOM DISINFECTANT, LAUNDRY DETERGENT, BLEACH, BUG KILLER, AIR FRESHENER AND MORE...

MANDY O'BRIEN AND DIONNA FORD

Ulysses Press

Published by
Ulysses Press
P.O. Box 3440
Berkeley, CA 94703
www.ulyssespress.com

ISBN: 978-1-61243-276-2
Library of Congress Catalog Number 2013947587

Printed in the United States by United Graphics Inc.

10 9 8 7 6 5 4 3 2 1

Acquisitions editor: Katherine Furman
Project editor: Alice Riegert
Editor: Kathy Kaiser
Proofreader: Lauren Harrison
Front cover design: Double R Design
Front image: © Aaron Amat/shutterstock.com
Interior design and layout: what!design @ whatweb.com
Index: Sayre Van Young

Distributed by Publishers Group West

NOTE TO READERS: This book has been written and published strictly for informational purposes. The authors and publisher are providing you with information in this work so that you can have the knowledge and can choose, at your own risk, to act on that knowledge. The views and opinions expressed in this book are solely those of the authors and do not necessarily represent those of Ulysses Press and/or the publisher.

This book is independently authored and published and no sponsorship or endorsement of this book by, and no affiliation with, any trademarked brands or other products mentioned within is claimed or suggested. All trademarks that appear in this book belong to their respective owners and are used here for informational purposes only.

We dedicate this book to everyone who is trying to make their home a safer, less toxic space.

TABLE OF CONTENTS

INTRODUCTION: SIMPLE STEPS

Modern society has witnessed a boom in health-conscious, environmentally aware individuals, but one vital component of overall well-being is often overlooked: our homes. Many of the products we use to keep our homes clean are toxic to the environment and our families. Marketers would have us believe that, in order to clean the right way, we must purchase an array of expensive products. This just isn't true. Making cleaning supplies with natural ingredients is not only possible, it can be done for much less money than you would pay for commercial products in the store. And while the idea of making everything from scratch may seem daunting, we hope to help you discover that making your own nontoxic cleaning supplies is actually quite easy.

Our journey to toxin-free homes began as we balanced being busy moms with our desire to make the healthiest choices for our families. Like many consumers, we feel a responsibility to leave the Earth a little bit better than we found it, but we also grapple with the daily reality of striking a balance between spending time with our families, doings things we enjoy, and taking care of ever-present to-do lists. We were both pleasantly surprised to find that creating homemade natural cleaning products was affordable, surprisingly fun, and shockingly simple. The key for each of us was gradually implementing reasonable changes that worked for our individual situations.

Ask those who live a comparatively green lifestyle, and they will all tell you the same thing: They didn't make their changes overnight. They made changes as they could, when they could. Living a healthier lifestyle is a continuum. There is no right way to make changes. Every individual and every family will have different needs and different capabilities.

With this in mind, we've divided our ideas and recipes into sections that felt natural to us when we thought about how we clean and manage our own homes. You can pick any one section and implement small changes to your routine at any time. For inspiration, we have included pop-out boxes in each chapter with simple changes to help you on your way; look for the simple step symbol ▃. Before you delve into the recipes, though, take a look at the next few sections, where we explain some of the science behind the products and processes used to clean in and around your home.

We hope that what you take away from this book is an understanding that living a less toxic life is an attainable goal. We hope that you will be inspired to take steps in whatever form works for you right now. Some of the changes may not seem practical to you now, but they might be a better fit later on. It's about simple steps.

THE BIG SECRET

The big secret of natural cleaning is this: You don't need the fancy bottles and the hard-to-pronounce substances inside them. A few essential supplies will have you cleaning naturally, healthily, and at a fraction of the cost of toxic commercial products. Hygiene doesn't have to come with a great cost to our wallets, and it shouldn't come with a cost to our health or to the environment.

> You don't need an arsenal of supplies to get your home clean. Use soap, water, and a few other basic ingredients, and keep your home toxin-free.

We are going to list some amazing, easy-to-make recipes to clean your home inside and out, but for the most part, you don't need all of these. If you are just starting out, choose some of the easy recipes and get started on healthier cleaning. As you watch your home get clean without the toxic chemicals, feel free to branch out with some of our other recipes or with your own concoctions. Before long, you will be adding essential oils to lift your mood and brighten your atmosphere. Many of our recipes are here to make cleaning more enjoyable. If you find it stressful to contemplate or use so many recipes, cut back to the basics. Making simple changes means just that: keeping it simple.

HOW SAFE ARE YOUR CLEANING SUPPLIES?

As consumers, we have been led to believe that by the time products make it to the market, they have been thoroughly tested and proven safe. We pick up packaging and read labels before buying, just to know what is in the product. But what does a label really tell us? Even for those of us who are trying to make better purchases, we really can't define what is safe from a product's label alone. Labels can be misleading. What we think we know about a product may not be true. How safe are your cleaning supplies, really?

VOLUNTARY DISCLOSURE

Do you know that companies that manufacture cleaning products for household and commercial use are not obligated to list the ingredients in their products? In 2010, the Consumer Specialty Products Association, the American Cleaning Institute, and the

Canadian Consumer Specialty Products Association created a Voluntary Disclosure Program. Developed by trade organizations whose primary purpose is to represent the interests of the companies engaged in the manufacture, formulation, distribution, and sale of household and institutional cleaning products, the Voluntary Disclosure Program leaves disclosure of ingredients at the sole discretion of the manufacturers. For those that do label their products, many ingredients might not make it to the label, making informed consumerism difficult.

> According to the National Research Council, no toxic information is available for more than 80 percent of the chemicals in everyday-use products.
>
> Only 1 percent of toxins are required to be listed on labels, because companies classify their formulas as "trade secrets."
>
> — Lorie Dwornick, researcher, educator, and activist, 2002

Companies are not obligated to disclose certain ingredients in commercial products, including expected contaminants (many of which are toxic to human and environmental health) and fragrances (which can include chemicals known to be allergens or hormone disruptors).[1] Even information listed on Material Safety Data Sheets can be incomplete. Manufacturers are required to list chemicals that are strongly linked to serious toxicity only if they account for more than 1 percent of the product's weight. They are required to list suspected carcinogens only if they account for more than 0.1 percent of the product's weight. Manufacturers can also legally omit hazardous chemicals if they are part of a mixture the company claims to be a trade secret.[2]

Manufacturers may choose to leave off any ingredients they wish in order to better market their product. In an age of information, we are ignorant of many of the ingredients used to clean our homes, businesses, and places of employment.

BIOACCUMULATION

Exposure alone isn't the only risk when it comes to toxic chemicals in your home. Although environmental levels may indicate low exposure with little risk, this doesn't speak to chemicals that exhibit bioaccumulation. Many toxic compounds, over time and with repeated exposure, can cause elevated levels of toxins in humans and other organisms. When looking at your health and the health of your family, be cognizant of the big picture. Something that may not cause problems today may, in time, cause long-term problems. Inform yourself. Protect yourself.

The term "mad as a hatter" refers to poisoning in the workplace. At one time, workers used mercury in the process of stiffening felt hats, which formed methyl mercury, a lipid-soluble compound that can accumulate in the brain, resulting in mercury poisoning.

CARCINOGENS

A carcinogen is a substance that is directly involved in causing cancer. Carcinogens are either genotoxic or nongenotoxic. Genotoxins cause irreversible damage or mutations by binding to DNA.[3] Nongenotoxins do not affect DNA directly but instead disrupt cellular metabolic processes and promote cancerous growth.[4]

Cocarcinogens do not cause cancer on their own. Instead, they promote the cancer-causing activities of carcinogens. Procarcinogens are precursors to carcinogens. These compounds are not carcinogens, but their environment can produce chemical reactions that turn them into carcinogens. Products you are using may not be carcinogenic on their own, but when used in your home and inhaled or absorbed through your family's skin, they may turn into carcinogenic compounds.

ENDOCRINE DISRUPTORS

Endocrine disruptors are chemicals that interfere with the endocrine system in mammals, resulting in changes in hormone production. The endocrine system consists of glands that secrete hormones and receptors that detect and react with the resultant hormones. Hormones work in very small doses. Because of this, low-level exposures to endocrine disruptors can have serious adverse effects.[5] Endocrine issues can result in cancerous tumors, birth defects, and other developmental disorders. Any bodily system controlled by hormones may be detrimentally altered by endocrine disruptors.

The first step in removing endocrine disruptors from the environment and our bodies is to discontinue production and use.[6] More difficult is the removal of hormone-disrupting chemicals, as they tend to be persistent, existing in the environment long after production has ceased. Some persistent organic compounds, such as polychlorinated biphenyl (PCB), dichlorodiphenyltrichloroethane (DDT), and polybrominated diphenyl ethers (PBDE), accumulate in aquatic sediment, where they continue to make their way into the food chain. The

Environmental Protection Agency (EPA) has outlined processes for cleaning heavily polluted areas in their Green Remediation program,[7] including the use of naturally occurring microbes that can degrade PCB in contaminated areas.[8] A few case studies regarding the cleanup of volatile organic compounds (VOCs) and PCBs in wetlands using native plants look promising,[9] but damage to organisms cannot be undone. It is easier to eliminate the use of damaging chemicals than to attempt to fix the damage that is already done.

NEUROTOXINS

Neurotoxins, strictly speaking, adversely affect the function of developing and mature nervous tissue.[10] This can result in widespread damage to the central nervous system, resulting in physical ailments, such as epilepsy,[11] or issues with brain function, such as dementia or memory impairment,[12] among others.

For thousands of years, humans have been incidentally exposed to neurotoxins. In the Roman Empire, lead poisoning was a significant problem. Citizens boiled sour wine in lead pans to sweeten it, generating lead acetate, which was referred to as "sugar of lead."[13] Neurotoxins also exist in nature. The venom of many animals includes neurotoxins, which target specific metabolic functions in order to immobilize prey for consumption or as a form of protection against other animals.[14] Advances in neuroscience have revealed not only the pathways in which neurotoxins work within our bodies, but also the potential hazards of using neurotoxic chemicals and compounds. From heavy metals to artificial chemicals, awareness of the hazards allows us to make informed choices regarding our lives.

VOLATILE ORGANIC COMPOUNDS (VOCS)

Volatile organic compounds (VOCs) from cleaning supplies used in residences have been measured at levels greater than 100 times higher than those found outdoors and can exceed safety limits established for industrial facilities.[15, 16] Although the discharge of VOCs in sewage treatment and storm water disposal is regulated as hazardous waste, there is no such regulation of VOCs released into nonindustrial indoor air.[17]

CAUTIONARY LABELS

Some products have cautionary labels, but the labels themselves can be confusing if you don't know what they mean.

Danger means one taste to one teaspoon is fatal to an adult.

Warning means one teaspoon to one ounce may be harmful or fatal to an adult.

Caution means one ounce to one pint may be harmful or fatal to an adult.

GREENWASHING: HOW COMPANIES FOOL CONSUMERS INTO BUYING "GREEN" PRODUCTS

What is greenwashing? Greenwashing occurs when companies "spend more time and money advertising that they are green than on actually putting into place environmentally friendly practices."[18] The term *greenwashing* was coined in 1986 to describe "the hotel industry's practice of placing placards in each bathroom encouraging guests to help the hotel conserve water by reusing their towels."[19] In reality, the reuse of towels does more to save the hotel chain money on laundry than it does to save the environment.

Most of the cleaning products sold in the green cleaning movement's infancy were either prohibitively expensive, ineffective, or "little more than colored water."[20] Many companies wanted consumers to think that natural products didn't work so that they would continue buying commercial products. Bottles of green colored water labeled as "all natural" or "green cleaners" had consumers believing that the heavy-duty cleaners were necessary to clean their homes when these products failed to work effectively.

As consumer interest in going green has grown, household cleaning product companies have jumped on the green bandwagon. But their vision of green often involves money, not the environment. Labeling a product "natural" or "environmentally friendly" is sometimes no more than a marketing ploy.

How do companies greenwash? One way is by making claims on labels that are false or misleading. Here are a few examples:

NONTOXIC: Although the Consumer Product Safety Commission (CPSC) requires that certain hazardous household products be labeled as toxic, there is no defined standard for the use of the term on consumer goods. As long as a product does not contain an ingredient specifically listed by the CPSC as toxic, a company may label their product as nontoxic. Unfortunately, the "nontoxic" label does not necessarily mean a product is safe.

NATURAL: The United States Department of Agriculture (USDA) dictates that meat and poultry carrying a "natural" label must not contain any artificial flavoring, coloring, chemical preservatives, or artificial ingredients. But when it comes to any other product on the market, just as with the nontoxic label, there is no set definition of what constitutes "natural." Manufacturers may add a "natural" label to their products at their sole discretion.

ENVIRONMENTALLY FRIENDLY: An "environmentally friendly" label is vague and impossible to prove or disprove. As with many other labels, there is no standard definition of the term, and no organization verifies claims of environmental friendliness.

BIODEGRADABLE: In order for a product to be labeled as biodegradable, Federal Trade Commission (FTC) guidelines state that it must degrade when exposed to air, moisture, bacteria, or other organisms, and it must return to the natural environment in a reasonably short time when properly disposed of. There is no standard or oversight for the label; manufacturers may label products "biodegradable" at their discretion.

Only items labeled as "certified biodegradable" have been certified by SCS Global Services. Based on peer-reviewed scientific studies and information provided by manufacturers, SCS determines whether the products will biodegrade after disposal, the toxicity of resulting components on aquatic life, the presence of phosphates or other compounds that will contribute to eutrophication, and whether the ingredients will cause displacement of other harmful substances already present in the environment. SCS has stringent certification guidelines and will not certify products if the ingredients are unlikely to degrade under practical conditions. SCS reviews the use of the SCS logo on labels and marketing materials to ensure proper usage. Without this certification, the label is meaningless.

Using misleading labels is only one way companies greenwash their products. Other sins of greenwashing include making unsubstantiated green claims, proclaiming irrelevant benefits, and glossing over a product's cumulative environmental impact. A 2010 survey by TerraChoice, an independent testing and certification organization, found that of 5,296 products examined, only 265 of them were really as green as they claimed. A whopping 95 percent of companies in their study were guilty of greenwashing.[21]

HOW COMPANIES GREENWASH CLEANING PRODUCTS, AND WHY THEY GET AWAY WITH IT

Going green is good business, and companies are eager to cash in on consumers' concern for the environment. Proclaiming a company to be eco-friendly is such an effective marketing practice that TerraChoice found a 73 percent increase in products being sold as green in a single year.[22]

How are companies marketing to people who are looking for safer, healthier alternatives but failing to deliver? In an initiative to uncover the truth about the toxic chemicals in ordinary household cleaners, independent scientists with the Environmental Working Group (EWG) reviewed more than 2,000 cleaning products.[23] The results are eye-opening — even among cleaners sold as green or natural, EWG reports that many contain chemicals that can cause serious damage to unsuspecting users. Following are a few examples of how cleaning products are greenwashed, several of which were revealed in the EWG Cleaners Database Hall of Shame:

Simple Green Concentrated All-Purpose Cleaner, which is labeled as "nontoxic" and "biodegradable," is sold in a spray bottle. The spray nozzle implies that the purchaser may use the product without diluting it, and there are no conspicuous directions on the bottle to the contrary. But the company's website advises users to significantly dilute the cleaner, even for heavy-duty cleaning. Spraying it directly on a surface without dilution puts the user at a greater risk of injury. And although this Simple Green cleaner is marketed as nontoxic, it contains a solvent that damages red blood cells and is an eye irritant (2-butoxyethanol), as well as a blend of alcohol ethoxylate surfactants, some of which are banned in the European Union.

Many oven cleaners, including certain Great Value, CVS, and Easy-Off products, "contain substantial amounts of sodium or potassium hydroxide, meant to dissolve crusty, baked-on gunk. These chemicals can also burn skin, lungs, and eyes." Despite this, these products are sold in aerosol form and can easily be inhaled or absorbed through the skin.

Comet Disinfectant Cleanser Powder contains as many as 146 chemicals, some of which may cause cancer, asthma, or reproductive disorders. Of course, Comet does not list several of the most toxic chemicals on the label, including formaldehyde, benzene, chloroform, and toluene.

Countless cleaning products use symbols (for example, the three arrows in a circle to show that a container is recyclable), colors (green!), pictures (leaves and trees are popular), and other marketing tools that aim to represent the green movement. For example, Dawn markets a dish soap with a label that includes the words "Dawn saves wildlife" alongside pictures of baby animals, because Dawn donates its antimicrobial soap to efforts to clean up animals after oil spills, and it donates money to other wildlife rescue efforts. But Dawn's dish soap contains triclosan, a synthetic, antimicrobial agent shown to be highly toxic to certain aquatic life and detrimental to aquatic ecosystems.[24]

As you can see, there are countless ways companies can greenwash cleaning products, in part because the industry remains largely unregulated. Because companies are not required by the federal government to list all ingredients on cleaning products, it is up to you to find out what chemicals you are spraying in your house and pouring down your drains. The EPA merely requires companies to list ingredients that are active disinfectants or are known carcinogens. The problem is that the EPA does not require companies to test ingredients, so relatively few ingredients are known to be harmful.[25]

In an ideal world, manufacturers would be required to prove that the chemicals in their cleaners were safe. But in the real world, under the Toxic Substances Control Act (TSCA), the burden is on the EPA to prove a chemical is unsafe.[26] In fact, since the TSCA was passed in 1976, the "EPA has required testing of less than 1 percent of the chemicals in commerce."[27] And of the approximately 85,000 chemicals in commerce today, the National Toxicology Program (a federal agency tasked with evaluating agents of public health concern) has the resources to test only ten to twenty of those chemicals each year.[28] For the vast majority of the chemicals we are exposed to every day, there is little to no data on any health or environmental risk — short-term or long-term.

This lack of federal oversight makes it difficult for the government to crack down on greenwashing. But under Section 5 of the Federal

Trade Commission Act, the Federal Trade Commission (FTC) has the authority to bring charges against businesses that make deceptive claims about their products. In other words, although the FTC cannot prevent a company from using toxic chemicals, it can prevent some of the greenwashing.

The FTC published "Green Guides" to help businesses make truthful and nondeceptive claims about any purported environmental benefits of their products. Though the guides are not enforceable rules or regulations, the FTC "can take action under the FTC Act if a marketer makes an environmental claim inconsistent with the guides."[29]

The "Green Guides" address many of the sins of greenwashing. Section 260.4, for example, advises that it is deceptive for a company to make a claim of a general environmental benefit. Such would be the case if a cleaning product's brand name were Eco-Friendly, because the company would not be able to substantiate that its entire line was indeed eco-friendly.

Although companies are not allowed to make broad claims of environmental friendliness, the FTC does not regulate the use of the words *natural* or *organic*. In fact, EWG reports that there are no federal regulations that require "cleaning supplies advertised as 'natural' or 'organic' [to] actually live up to [claims of being safer]. Products bearing the U.S. Department of Agriculture's 'organic' seal contain ingredients from plants grown without artificial pesticides and fertilizers, but these products aren't automatically safer for you or the environment." Some cleaning products, for example, contain natural, organic ingredients like "linalool, eugenol, and limonene, which are natural components of essential oils that can trigger allergic reactions." Moreover, even if you do find products that have received seals of approval or "green" certifications, it can be hard to figure out what certification programs are trustworthy.[30]

WHAT YOU CAN DO

Manufacturers, and the organizations that work to protect their interests, aren't going to make changes. If, as consumers, you want the right to know exactly what is in the products you purchase and to change the toxicity of products used all around you, you need to ask for and demand change.

Due to pressure from constituents, several states have passed laws that only permit green cleaning supplies to be used in state buildings and schools.

Use your money to make a stand. Scrutinize labels and ingredients. Research the products you buy. Purchase only from companies that list all of the ingredients on their products' packaging. Try making your cleaning products at home, using less toxic ingredients.

When you do have a concern about a commercial product, contact the company and express your concerns about their ingredients and/or marketing practices.

Alert professional and governmental organizations charged with monitoring companies, including the Better Business Bureau, Greenpeace, and the Federal Trade Commission.

Ask the places you frequent — schools, daycares, workplaces, or community centers — to change the products they use to safer, less toxic supplies.

Support state and federal efforts requiring manufacturers to disclose all ingredients on labels. Consumers need that information available to them at the point of purchase.

Support the reform of the outdated federal system's methods of evaluating and approving toxic chemicals.

NONTOXIC AND GREEN CLEANING MYTHS

Natural, or green, cleaning often gets a bad rap. People think green cleaning is not as effective or harder or just too much trouble. These ideas are perpetuated by companies out to make a dollar. They don't want you to realize that you don't need their products to get a sparkling clean toilet, counters you could eat off of, or a house that is clean from top to bottom. Their bottom line is your bottom dollar.

MYTH: Commercially available products have been thoroughly tested and proven safe. Manufacturers would be required to disclose if their products weren't safe.

Manufacturers would like you to believe this is true. The fact is, you may not even know all of the ingredients in the products you are buying. Due to lax laws regarding cleaning products, companies are not required to list all ingredients on a product's label. And due to loopholes in the laws, many ingredients that have been proven unsafe are allowed to be used in cleaning products.

MYTH: Modern products are more effective than what previous generations used.

We have learned a lot over the years about cleanliness and hygiene. We have also learned a lot about how various chemicals whose use began generations ago are affecting us, either directly or through environmental changes. Many toxic chemicals, both natural and artificial, have been used throughout history. What is important is getting back to basics and using the appropriate products to clean thoroughly and efficiently without adding to the biological load on ourselves or the environment.

MYTH: If it smells stronger, it cleans better.

We tend to associate smells with experiences and ideas, but that smell wafting off your newly cleaned home isn't the smell of clean. Clean actually doesn't have a smell, as an item that is clean has had any offending dirt or bacteria removed. Any smells you associate with cleaning are actually the result of fragrances, commercially formulated or naturally occurring, in the products you use.

MYTH: Natural cleaners don't work as well as mainstream products.

Do green cleaners really work as well as mainstream products? Well, it depends on what you are using to clean with. Most of the time, you can't expect to swish some water over an area and have it be clean. That's not cleaning. That's getting something wet. However, if you use a natural cleaner that works, then it will work as good as, or even better than, mainstream products — without the harmful chemicals most mainstream products contain.

MYTH: Green cleaning won't kill germs.

Just because natural cleaners weren't concocted in a lab doesn't mean that they won't kill germs. Natural cleaning still uses science to get things clean. Studies have shown that cleaning a surface with a vinegar mixture, followed by hydrogen peroxide, actually does a better job of cleaning viruses and bacteria than bleach. Plus, you will not have to contend with the awful fumes or health issues associated with bleach.

MYTH: Nontoxic cleaners are more expensive than chemically based products.

Nontoxic cleaners on the market are generally priced competitively with their more mainstream counterparts. However, you don't need to pay even that much. In just a few minutes, you can make your own natural cleaning products for a fraction of what you pay at the store. By doing so, you will know for certain what ingredients are used and that the item is actually safe around your family, while saving money at the same time.

MYTH: Using nontoxic cleaners takes more energy.

Many people, even those who choose to use natural cleaners, often say that cleaning with natural cleaners just takes a little more elbow grease. Don't believe it. Clean smarter, not harder. Just as with mainstream products, you need to let the cleaner do its job. Don't expect to spray something and have the area become instantly clean. Dirtier areas may need a little more time. Spray, walk away, and come back in a few minutes. Use the right products for what you are trying to accomplish, and let the products work for you.

MYTH: Making your own cleaning supplies takes a lot of time.

If making your own cleaning supplies took a lot of time, people wouldn't turn to this method, either to cut down on toxic chemicals or to save money. The fact is, making up a solution generally takes only a few seconds to a few minutes. Grab some supplies — most of which you probably already have — and start cleaning!

MYTH: Homemade cleaning supplies use hard-to-find ingredients.

Our great-grandmothers knew all about using what they had on hand to do the jobs they needed to do. They weren't scouring the stores for obscure ingredients to make complicated concoctions. Today's natural cleaning stays true to that. You will find most of the ingredients we list at your local grocery store. Essential oils, once made in kitchens from backyard herb gardens, can be found at your local natural foods store or through the Internet. One whiff and you'll be hooked on these wonderful-smelling oils with the power of plants to help fight bacteria, viruses, and more. But take only a tiny whiff — essential oils are potent!

MYTH: Even natural chemicals are chemicals, so it doesn't matter what you use.

This is only half-myth. It is true that everything in our lives is made of chemical elements and compounds. That doesn't make them all the same, though. Many chemicals, the very same ones sold in most commercial products for you to clean your home with, are toxic to your health. Making healthier choices is easy to do and helps reduce your overall exposure.

GETTING STARTED

Thinking about switching your cleaning products can be overwhelming. Many people are concerned that they will need to spend a lot of money buying ingredients or a lot of time finding the ingredients needed to make their own nontoxic products. Luckily, this isn't true. Most of the ingredients for all-natural cleaning are probably already in your kitchen. If not, you can easily find most of them at your local grocery store. Although you may want to branch out by adding more essential oils over time, rest assured that this isn't a requirement. We want to help you clean and make your home safer without spending a lot of time or money. There are more important things in life than cleaning.

BASIC CHEMISTRY

The business of manufacturing cleaning products relies on chemistry. Were you to look at the ingredient list of these products, you would read a list of chemical names, many of which might be unfamiliar. Cleaning does come down to chemistry and biology, but it is much simpler than manufacturers would have us believe.

Key components of cleaning products are acids and bases. Whether acidic or alkaline, cleaning products use acids and bases in order to neutralize stains and odors by bringing the pH of the item being cleaned back to 7. Acids have a pH less than 7 and bases have a pH greater than 7. This is also true for natural cleaning supplies. Some of the basic ingredients used are natural acids, such as white distilled vinegar or lemon juice, and natural bases, such as baking soda or borax.

The acidic nature of white distilled vinegar not only aids in its neutralization of stains and odors but also makes it an effective descaler of hard minerals. Only white distilled vinegar should be used in cleaning, as the coloration of apple cider vinegar may stain.

Forget the elbow grease and let your homemade cleaning supplies do the work.

The more alkaline products, while neutralizing, owe many of their cleaning properties to their status as minerals. The alkaline aspect cuts grease and other substances, such as wax, while the minerals work as a mild abrasive. Due to their abrasive nature, many homemade cleaning products have gotten a bad rap as requiring more elbow grease. However, mixed in the correct proportions, used in sufficient amounts, and given enough time to work their chemical magic, homemade cleaning products do not require any more work than their commercially made counterparts.

SOAP VS. DETERGENT

We tend to think of the terms *soap* and *detergent* as interchangeable. It is true that both are surfactants, compounds that easily mix together with grease and water in order to cut through grease to remove it. Chemically, these two items are very different. Detergents are generally made from petroleum distillates and have a much harsher impact on the environment. They contain toxic pollutants, which affect wildlife and humans alike.[31] The resulting products, made from nonrenewable resources, are not nearly as biodegradable as their natural soap alternatives. Soaps, on the other hand, are made primarily from saponins, the sudsing agents found in the roots of soapwort, soapberry, and yucca plants, and use natural oils, such as lard, olive oil, or coconut oil.

BASIC SUPPLIES

More than 32 million pounds of household cleaning products are poured down the drain each day nationwide. The toxic substances found in many of these are not adequately removed by sewage treatment plants. Guess what happens when these are returned to the rivers from which cities draw their drinking water?
— "Cleaning Without Toxic Chemicals," Spring 2002, CCA Newsletter Partners

A few basic ingredients are all you really need to start making your own cleaning supplies. We have included a list of some of these items, along with others that are included in the recipes that follow. Don't let the length of the list unnerve you; it is information, not a prescription. Take it slow. Use some of the basic ingredients you probably already have in your house. As you become more adventurous and want to try new recipes or add some natural scents, pick up additional items, as needed.

Grab these easy basics, which you probably already have at home, for your natural supplies: white vinegar, baking soda, and castile soap. Throw in some tea tree oil for extra cleaning power.

Baking Soda

Baking soda, also known as bicarbonate of soda, is an alkaline base that reacts with acids, including a majority of liquids. It exists in nature, including in the human body, where it is a major component of bile, used to neutralize stomach acid. Baking soda can be purchased in grocery and discount stores. Small boxes of baking soda are located with the baking supplies, and larger boxes or bags may be found near the laundry products.

White Distilled Vinegar

Due to its acidic nature and antibacterial effects, white distilled vinegar is used in many homemade cleaning products. Japanese researchers found that the antibacterial efficacy of vinegar directly increases with the temperature of the solution.[32] White distilled vinegar can be found at any grocery store.

Lemon Juice

This simple product is an acid commonly found at your local grocery store. Lemon juice kills mold,[33] cuts through grease, and leaves a streak-free shine.

Borax

Also known as sodium borate, sodium tetraborate decahydrate, or disodium tetraborate, borax is a naturally occurring mineral and a salt of boric acid. Look for borax in the laundry detergent aisle of your grocery store. Note: Borax should be used with care, as exposure to borax dust can cause respiratory and skin irritations.[34] Special care should be taken when pouring the product to avoid inhaling the fine particles.

Soap

As we tell our children when they wash their hands, "Use soap." Soap does wonders for cleaning many items around your home, including you! Confused about which soap to use? As previously mentioned, true soap is made from natural saponins, rather than petroleum products, and can be made from a wide variety of fats. Products labeled "soap" are most likely made from animal products. Those labeled "castile soap" are made from plant-derived fats, such as olive oil, coconut oil, or almond oil. Most true soaps currently on the market are castile soaps. They have become very popular and can be found almost anywhere, from local grocery and chain stores to natural foods stores and co-ops. Some castile soaps are scented with essential oils. You can also use plain castile soap or add your own scents using essential oils.

Hydrogen Peroxide

Hydrogen peroxide (H_2O_2) is a strong oxidizer, making it useful as a bleaching and cleaning aid. It occurs naturally in low concentrations in nature as a by-product of oxidative metabolism in living organisms. Hydrogen peroxide is considered a highly reactive oxygen species. Concentrated hydrogen peroxide solutions are sometimes used as a rocket propellant.[35] Diluted 3 percent solutions are sold to the public and used for cleaning. Look for hydrogen peroxide in the personal care section of your grocery store.

Vodka

Available anywhere alcohol is sold, vodka does contain alcohol, but this alcohol is ethanol, which is less toxic than many other alcohols, such as methanol,[36] ethylene glycol,[37] or diethylene glycol,[38] often found in cleaning products. The majority of the recipes in this book do not use alcohol, but a few do. Vodka can work well for making extracts or in sprays where you need the carrier base to evaporate quickly.

Washing Soda

Sometimes referred to as soda ash or soda crystals, washing soda is actually sodium carbonate. Domestically, washing soda is used as a water softener, preventing magnesium and calcium ions in hard water from bonding with detergent. It is often used to remove tough stains, such as grease or oil, and as a descaling agent in the boilers of various household appliances. Find washing soda in the laundry section of your favorite store.

Water

Water is a key component in many of the recipes in this book. Water is readily available, but not all forms are equal. Be aware of your water and your needs. Filtered water may be more appropriate for some applications, especially if your tap water is considered to be hard from heavy mineral concentrations or when there is risk of biological contamination. Plain tap water is sufficient for many cleaning needs. Choose what works best for you.

Witch Hazel

Witch hazel is an extract made from alcohol and the leaves of the witch hazel shrub (*Hamamelis virginiana*). Check the personal care section of your grocery store for this item, which is sold as a product in its own right and is also found as an ingredient in many personal care products.

ESSENTIAL OILS

Essential oils are natural oils extracted from plants, including trees, shrubs, herbs, grasses, and flowers, through steam distillation, expression, or solvent extraction. Unlike many artificial chemicals, which can accumulate in our bodies or in the environment, the plant nature of essential oils results in a product that breaks down readily. Research regarding essential oils has confirmed their various applications, including antibacterial, antifungal, and antiviral, depending on the essential oil.[40]

Although essential oils are all natural, use them with caution. As most essential oils are very concentrated, they should be diluted using a carrier oil as a base. In this book, we discuss essential oils used topically or in cleaning solutions; please do your own research for using essential oils internally, as some can cause harm. Keep them out of reach of children or pets. Additionally, many essential oils are photosensitive and heat-sensitive and are best stored in a cool, dark place.

We have included a list of the essential oils we used in the recipes in this book. You do not need to purchase or use all of these oils. If you find a recipe you would like to try, check out any suggested essential oils and decide what will work for you. Some oils will appeal to some individuals more than others, and many have similar properties. Experiment. Go slow. Before long, you will enjoy using essential oils in many of your home applications.

To learn more about essential oils and how you might use them for other applications besides cleaning, check out Valerie Ann Worwood's book *The Complete Book of Essential Oils and Aromatherapy*.[41] It is an excellent beginner's book about essential oils.

BERGAMOT OIL (*CITRUS BERGAMIA*): This essential oil, cold-pressed from bergamot orange fruit rind, has been used by the fragrance industry since the early 1700s. Earl Grey tea is a black tea containing bergamot oil.[42] *Analgesic, antibacterial, antiseptic, antispasmodic.*

CATNIP OIL (*NEPETA CATARIA*): Used primarily in natural insect repellents,[43] catnip oil is most notable for its ability to excite cats; it also acts as a sedative for humans and is used in many blends, including many used to treat insect bites or to repel insects. It is related to mint and has a characteristic mintlike aroma. *Anesthetic, anti-inflammatory.*

CEDAR OIL (*JUNIPERUS VIRGINIANA*): Used in many body care products, cedar also has insect-repelling qualities. The woodsy scent blends well with many other wood-derived essential oils. *Antiseptic, antispasmodic.*

CHAMOMILE OIL, GERMAN (*MATRICARIA RECUTITA*): Used in many body care and health products, chamomile also has properties that make it well suited for cleaning, including its natural antibacterial and antifungal properties.[44, 45] Chamomile has a subtle scent and works well blended with many other essential oils. *Analgesic, antibacterial, antifungal, anti-inflammatory, antispasmodic.*

Use the natural power of essential oils to clean and disinfect your home. Start with tea tree oil and lavender oil for their antimicrobial, antifungal, and antibacterial properties.

CHAMOMILE OIL, ROMAN (*ANTHEMIS NOBILIS*): Roman chamomile has a sweet scent with a hint of apples. It is used in many natural body care products and also works well in aromatherapy. Blends with Roman chamomile are an excellent choice for air diffusion. *Analgesic, antibacterial, anti-inflammatory, antimicrobial, antiseptic, antispasmodic.*

CINNAMON BARK OIL (*CINNAMOMUM ZEYLANICUM*): The smell of cinnamon may have your olfactory senses in overdrive, remembering falls and winters, but the oil from the trees' bark has many beneficial properties.[46, 47] It is perfect for disinfecting surfaces and the air to help keep your family healthy, especially during those seasons when colds and other illnesses are at their peak.[48, 49] *Analgesic, antibacterial, antifungal, anti-inflammatory, antimicrobial, antispasmodic, disinfectant.*

CITRONELLA OIL (*CYMBOPOGON WINTERIANUS*): Best known for its use in natural insect repellents, citronella can also be used to treat insect bites and is used in some home remedies.[50] It blends well with woodsy and citrus essential oils. *Analgesic, antibacterial, antifungal, antiseptic, antispasmodic.*

CLOVE BUD OIL (*SYZYGIUM AROMATICUM*): Clove oil has long been used in natural medicines. Due to its effect on small organisms, such as bacteria and viruses,[51] clove bud oil used in blends works well to help rid your home of potential illness. The oil blends well with many other essential oils. *Analgesic, antibacterial, anticlotting, antifungal, anti-inflammatory, antimicrobial, antiseptic, antispasmodic, antiviral.*

CYPRESS OIL (*CUPRESUS SEMPERIVENS*): This essential oil traditionally has been associated with religious cleansing rituals and has been used in burial processes to protect workers. Cypress oil makes an excellent substitute for pine essential oil when used in cleaning. *Antibacterial, anti-inflammatory, antiseptic, antispasmodic.*

EUCALYPTUS OIL (*EUCALYPTUS GLOBULUS*): Eucalyptus has been used in natural medicines for millennia and is one of the oldest medicines native to Australia. Many commercial products still use eucalyptus to combat viral and bacterial illness.[52] Eucalyptus oil is excellent for use in disinfecting or in a diffuser to cleanse the air during times of potential illness. *Analgesic, antibacterial, antifungal, antiseptic, antispasmodic, antiviral, disinfectant.*

GERANIUM (ROSE) OIL (*PELARGONIUM GRAVEOLONS*): Used in many natural medicines to help heal and prevent illness, geranium (rose) oil blends well with many other essential oils in homemade cleaning products, in body care products, and in diffusers to cleanse the air. *Analgesic, antibacterial, anti-inflammatory, antiseptic.*

GRAPEFRUIT OIL (*CITRUS PARADISE*): Grapefruit has a lovely citrus scent that blends well with other essential oils. *Antibacterial, antiseptic.*

Add a little natural scent to your cleaning with essential oils. Consider lemon oil for a fresh scent, ylang ylang oil for a floral scent, or pine oil for a woodsy scent.

JASMINE ABSOLUTE OIL (*JASMINUM GRANDIFLORUM*): Jasmine absolute oil has a long history in the perfume industry. It blends well with many other essential oils and has a pleasant floral scent. *Analgesic, anti-inflammatory.*

JUNIPER OIL (*JUNIPERUS COMMUNIS*): This essential oil works well in many earthy blends. Juniper oil is traditionally used to purify the air but is also used in some insect repellents. *Analgesic, antimicrobial, antiseptic, antispasmodic.*

LAVENDER OIL (*LAVENDULA ANGUSTIFOLIA*): Lavender oil is one of the most versatile of essential oils for the home. Used as a natural antiseptic and pain reliever for minor burns, bites, and stings,[53] it is also known for its ability to help aid in relaxation[54] and anxiety reduction.[55] *Analgesic, antibacterial, antifungal, anti-inflammatory, antimicrobial, antiseptic, antispasmodic.*

LEMON BALM OIL (*MELISSA OFFICINALIS*): Lemon balm is known for its calming effects and ability to help repel insects.[56] Also used in many natural medicines, the essential oil has cleaning properties that make it a welcome addition to any nontoxic home.[57] *Antibacterial, antihistaminic, anti-inflammatory, antiseptic, antispasmodic, antiviral.*

LEMON EUCALYPTUS OIL (*EUCALYPTUS CITRIODORA*): Commercially, lemon eucalyptus has been used primarily in the fragrance industry, but new research has made it popular in natural bug repellents. Lemon eucalyptus essential oil has a light, lemony scent and blends well with other essential oils for cleaning. *Antibacterial, antifungal, antiseptic, antiviral.*

LEMON OIL (*CITRUS LIMON*): Lemons have long been used for their medicinal contributions as well as their cleaning abilities. Lemon oil is a staple in many homes and makes a welcome addition to natural cleaning. *Antibacterial, antifungal, anti-inflammatory, antimicrobial, antiseptic, antispasmodic.*

LEMONGRASS OIL (*CYMBOPOGON FLEXUOSUS*): Lemongrass essential oil has received recent exposure for its use as a natural insect repellent. The lemony aroma, along with its other properties, makes it a valuable addition to natural cleaning supplies.[58] *Analgesic, antifungal, anti-inflammatory, antimicrobial, antiparasitic, antiseptic, antiviral.*

LIME OIL (*CITRUS AURANTIFOLIA*): Similar to lemon essential oil in its qualities, lime essential oil makes a nice variation to enliven your homemade cleaning supplies. *Antibacterial, antiseptic, antispasmodic, antiviral.*

ORANGE (SWEET) OIL (*CITRUS SINENSIS*): Sweet orange essential oil contains approximately 90 percent limonene,[59] which is a natural ingredient found in many commercial household cleaners. It makes an excellent addition to your homemade cleaning supplies. *Antibacterial, anticoagulant, antifungal, anti-inflammatory, antiseptic, antispasmodic.*

According to Valerie Ann Worwood in her book, *The Complete Book of Essential Oils and Aromatherapy*, oregano essential oil is twenty-six times as powerful an antiseptic as phenols, which are found in many commercial cleaners.[39]

OREGANO OIL (*ORIGANUM VULGARE*): While the plant itself is generally used for culinary purposes, oregano essential oil works effectively against microbes, fungi, bacteria, and parasites.[60] It blends especially well with wood-based essential oils. *Analgesic, antibacterial, antifungal, antimicrobial, antiparasitic, antiseptic, antispasmodic.*

PEPPERMINT OIL (*MENTHA PIPERITA*): Well known for its culinary and medical uses, peppermint essential oil also makes a great addition to the home cleaning basket due to its many beneficial properties.[61] It blends well with a variety of other essential oils, adding a minty aroma. *Analgesic, antibacterial, antifungal, antiinflammatory, antimicrobial, antiseptic, antispasmodic.*

PINE OIL (*PINUS SYLVESTRIS*): Pine essential oil is well known for its woodsy aroma. Many commercial products have mimicked the scent with artificial fragrance. Consider using this natural essential oil to boost your cleaning. *Analgesic, antibacterial, antifungal, anti-inflammatory, antimicrobial, antiseptic, antiviral.*

SAGE OIL (*SALVIA OFFICINALIS*): Sage plants have long been used both for food and for their healing properties (*Salvia* means "health"). The species name, *Salvia officinalis*, is derived from the word *officina*, the traditional storeroom of a monastery where herbs and medicines were stored.[62, 63] Pregnant and lactating women should take care when handling or ingesting sage essential oil or sage herbs. *Antibacterial, anti-inflammatory, antimicrobial, antiseptic, antispasmodic.*

SPEARMINT OIL (*MENTHA SPICATA*): Well known for its use in cooking and medicine, spearmint essential oil's other natural properties, along with its milder nature when compared with peppermint essential oil, make it an excellent alternative for cleaning.[64] *Analgesic, anesthetic, antibacterial, anti-inflammatory, antiseptic, antispasmodic.*

TEA TREE OIL (*MELALEUCA ALTERNIFOLIA*): Tea tree oil is a staple in the home of most everyone cleaning with natural products. Its natural properties make it perfect for use in a wide variety of cleaning applications.[65, 66] It blends well with many other essential oils and works well on its own when diluted. If you are looking for one all-purpose essential oil to buy first and get you started with natural cleaning products, this is the one to use for disinfecting the home. Its distinctive scent is an acquired taste for some. *Analgesic, antibacterial, antifungal, anti-inflammatory, antimicrobial, antiparasitic, antiseptic, antiviral, disinfectant.*

During World War II, the Australian military issued every soldier a bottle of tea tree essential oil to treat infections.

THYME OIL (*THYMUS VULGARIS*): Medicinally, thyme essential oil is well known for its antiseptic and disinfectant properties.[67, 68] It blends well with a wide variety of other essential oils. *Anagelsic, antibacterial, antifungal, anti-inflammatory, antimicrobial, antioxidant, antiseptic, antispasmodic, antiviral, disinfectant.*

YLANG YLANG OIL (*CANANGA ODORATA*): Ylang ylang essential oil is known for its pleasant floral scent and has often been used by the fragrance industry. It blends well with many other essential oils and makes a fragrant and calming addition to a natural cleaning basket. *Antibacterial, antifungal, anti-inflammatory, antiseptic, antispasmodic.*

BASE OILS

When we discuss recipes that contain essential oils, we talk about base oils (also known as carrier oils). Base oils are exactly what they sound like: a base to which you add essential oils. Because essential oils are concentrated, they are often too strong to use on their own. A base oil dilutes the essential oils, making them safe to use; it also helps you to conserve your essential oils, making them easier on your budget. Why use more of a product than you need?

There are myriad options when it comes to base oils, but if you are just getting started, you can start out with oils you already have in your kitchen. There are no hard and fast rules about base oils. Feel free to substitute the base oils that fit your needs and your budget. The same oils you cook with can often be used to help clean dirt, protect against fading, hide imperfections, and more.

ALMOND OIL (SWEET) (*PRUNUS DULCIS*): Almond oil, derived from the commonly eaten tree nuts, makes a wonderful addition to your home. Not only can you use it to help clean your home in a nontoxic way, but it is excellent for your skin. You just may decide to replace some of your personal care items with homemade, nontoxic versions. When purchasing almond oil, make certain you buy sweet almond oil. This is the primary form you will find, but bitter almond oil sometimes still makes its way to the market. Bitter almond products, whether the oil or extract, contain an enzyme that, when combined with water, yields cyanide.[69] Almond oil is a light golden color. You can find it at your local natural foods store, online, or at some grocery stores.

For easy base oils, check out some of the oils you may already have in your kitchen, such as coconut oil, olive oil, or sunflower oil.

APRICOT KERNEL OIL (*PRUNUS ARMENIACA*): Used less often than almond oil, apricot oil, pressed from the kernels of apricots, shares many properties with almond oil. It is excellent for your skin and can be used in cleaning as well. Apricot kernel oil has a deeper gold color than almond oil. Check online or at your favorite natural foods store.

COCONUT OIL (*COCOS NUCIFERA*): Coconut oil makes a wonderful base for many household and personal care products. The oil is a white solid at lower temperatures, melting to a clear liquid when summer approaches and the temperature in your home begins to rise. When the temperatures drop back down, the coconut oil will resolidify. You will find both refined and unrefined versions on the market. The beneficial antibacterial and antimicrobial properties of coconuts are retained in unrefined coconut oil,[70, 71, 72] making this natural product an ideal cleaning agent and ingredient in other homemade concoctions. The virgin oil also retains the characteristic coconut aroma and a slight taste of coconut. This oil can be found online or at most grocery stores, natural foods stores, and discount stores.

JOJOBA OIL (*SIMMONDSIA CHINENSIS*): Jojoba oil is a bright golden color, well known for its use in skin care products. It is a very stable oil with a high wax concentration, making it useful as a wood polish. Although the oil is not used in cooking,[73] its qualities make it a natural for homemade applications. Check the oils or skin care sections of your natural foods store or the natural foods shelves of your grocery store.

OLIVE OIL (*OLEA EUROPAEA*): Olive oil can be found in almost every home. Known for its culinary applications, especially with Mediterranean recipes, olive oil is very stable. It is also a helpful addition to homemade household and personal care products.[74] Olive oil ranges from a slight olive color to a light brown color. This oil can be found at most grocery stores in a wide range of prices.

SUNFLOWER OIL (*HELIANTHUS ANNUUS*): Sunflower oil, a light amber in color, has many applications. It has some natural preservative properties for both food and personal care products and also can be used in homemade household recipes. Check for sunflower oil at your local natural foods store or the natural foods section of your local grocery store.

CONTAINERS

Reusing containers, such as spray bottles, from your current cleaning supplies might seem both economical and environmentally friendly, but this can be counterproductive. Chemical residues in the containers can contaminate your new supplies. Instead, dispose of chemical containers properly, and put your new products in different containers.

REPURPOSING

Certainly, you should be wary of repurposing containers that have held chemicals, but there are many other containers around your home that can be repurposed. Glass jars from food are safe and perfect containers for many of the items you make. Some other food containers, free of toxic chemicals, make wonderful containers for cleaning supplies, as well. Some even have tops that make sprinkling items, such as baking soda, easy. Look around. You might be surprised by how many containers you already have.

> Repurposing glass jars is an economical and environmentally friendly way to store your homemade cleaning supplies.

PURCHASING

You can find containers in every price range, but you don't have to spend a lot of money to store your new toxic-free cleaning supplies. BPA-free spray bottles can be purchased for as little as $1. You can find fancier, more expensive containers, but you definitely don't need to spend a lot of money.

LABELS

Whatever container you choose, label it well. Use stick-on labels, a permanent marker, or another method, and list the recipe on the label. Not only will you know exactly what is in the container, but when you come up with a blend you love, you will be able to re-create it.

OTHER SUPPLIES

The other recommended supplies are items you already have on hand. You may want a bucket and mop or brush for mopping. Rags or cloths can be used for everything from cleaning glass or countertops to washing dishes or scrubbing dirt (some of our favorite alternatives to paper towels follow).

> Replace many of your disposable cleaning products with simple cloth wipes (see page 240). They work for a wide variety of applications.

PAPER TOWELS

Americans use approximately 13 billion pounds of paper towels every year. Unlike some other paper products, paper towels are generally not recyclable. Not only do their thin fibers make recycling difficult, but they are often full of bacteria from wiping up spills and messes.[75]

If every American used just one fewer sheet each day, we would reduce our paper waste by 571,230,000 pounds in a year.[76] That's just one sheet per day; how many more millions of pounds of waste could we save by switching to cloth?

If 10,100 families of four replaced a year's worth of 180-sheet virgin fiber paper towels with 100 percent recycled paper towels, they would save 864,000 trees, 3.4 million cubic feet of landfill space (it would take 3,900 full garbage trucks to fill that space), and 354 million gallons of water.

Paper towel alternatives are easy and affordable. To dry your hands, invest in a pack of kitchen or other hand towels for the kitchen and bathroom. Hang one up at each sink, and wash every other day to keep germs away. If you have a larger family, consider having several available, including one at a lower level for younger members of the family.

Paper Towel Alternatives

Here are our favorite paper towel alternatives for wiping up spills and for cleaning:

REPURPOSED T-SHIRTS: To clean up spills around the house, repurpose your old 100 percent cotton T-shirts. Snip off the sleeves, cut straight across the neckline and waistline (unless you want those reinforced seams; either way is fine), and cut down the sides. Keep them in a basket near the sink, on or under your kitchen and bathroom counters, and in your hall closet or other cleaning supply cabinet. You'll always have cloths handy for messes, and you don't need to worry about keeping them pretty.

REPURPOSED TOWELS AND WASHCLOTHS: For messier cleanup jobs, have some old washcloths and cut-up towels nearby. No old towels? Stop by a thrift store. You will find a treasure trove of paper towel alternatives for a fraction of the cost. Old towels are also wonderful for pet messes or for preventing kid messes. Spread the towels out for snacks during movie night or particularly messy activities, such as painting.

CLOTH DIAPERS: New or used, cheap cotton flat or prefolded diapers are a great green alternative to paper towels. Their high absorbency makes them great for soaking up spills.

COMMERCIAL BAR OR SHOP TOWELS: If you want something really heavy-duty, consider investing in some industrial-strength shop towels. They're built to clean up grease and grime, and you'll recoup the cost when you stop buying paper towels. Most old towels will do the job just as well, though.

COMMERCIAL PAPER TOWEL ALTERNATIVES: From bamboo to cotton-and wood cellulose blends, paper towel alternatives come in many fabrics. They are washable and reusable, but they often have a much shorter life than regular old towels or repurposed cotton. One of the leading bamboo towels, for example, is supposed to last 10 days. Repurposed towels and cloths last decades!

Paper Towel Alternatives for Cleaning

Almost all of the alternatives mentioned previously are also great for cleaning, but when you want a lint-free shine, try one of the following.

FLOUR SACK TOWELS: Flour sack towels are 100 percent cotton, thin, and lint-free, making them perfect for cleaning and shining (but not so great for wiping up spills). Use them on glass or mirrors or anywhere you need a streak- and lint-free shine.

MICROFIBER TOWELS: Microfiber towels are absorbent and should not leave lint behind.

RECYCLED PAPER: If you want to remove the lint on your glass after cleaning with a regular towel, wipe it down with a piece of newspaper, junk mail, or a page ripped out of the phone book.

If you must use paper towels, then be mindful about what kind of paper towel you purchase. Look for a brand that uses 100 percent recycled fiber and manufactures with either totally chlorine-free (TCF) or processed chlorine-free (PCF) bleaching.[77] Instead of throwing paper towels away, place compost-safe used paper towels in your compost bin or check to see if your waste disposal provider accepts them in yard waste.

SIMPLE CLEANING

Today's stores are filled with a plethora of cleaning supplies to help you wage war on dirt and grime in your home. It's a good marketing strategy: The more products there are, the more you think you need to buy. Before long, in your quest to eradicate your home of dirt and germs, you find that you have an entire cabinet full of products, most of which are probably toxic. The result? In your attempts to create a healthier home, you are making your home anything but.

ANTIBACTERIAL PRODUCTS

Commercial antibacterial products have enjoyed an increase in sales in the past couple of decades as concerns about bacteria have grown. This is especially true of families with children. You can purchase everything from antibacterial soaps and cleaners to cutting boards, storage containers, mattresses, socks, toys, and more. But are antibacterial products as helpful as they claim — or even safe?

According to Rolf Halden, cofounding member of the Center for Water and Health at Johns Hopkins Bloomberg School of Public Health, triclosan and triclocarban are present in 60 percent of America's streams and rivers. High concentrations in soil, combined with pathogens from sewage, could be "a recipe for breeding antimicrobial resistance."

Unlike other cleansers, antibacterial products leave behind additional surface residues. This residue includes active chemicals that continue to kill some, but not all, of the bacteria on a surface. The remaining bacteria are stressed by the commercial product, causing a subpopulation to develop defense mechanisms.[78] In other words, the stronger bacteria, which are able to adapt to the chemicals, live and reproduce. Over time, the bacteria become resistant to various chemicals and antibiotics through genetic mutations; this results in cross-resistance.[79]

Triclosan and a similar chemical, triclocarban, are both widely used in antibacterial products. Triclosan has been found throughout the environment, from water and soil to the milk of mammals.[80] Triclosan may be an endocrine disruptor; it breaks down into dioxin, a known carcinogen.

Both the American Medical Association and the Centers for Disease Control and Prevention point out that consumers don't need antibacterial soaps in their homes to stay clean. The U.S. Food and Drug Administration (FDA) has published reports saying that there are no medical studies showing a decline in infection rates for consumers who use antibacterial products. What should you do instead?

- **WASH YOUR HANDS THOROUGHLY WITH SOAP AND HOT WATER THROUGHOUT THE DAY.** Wash long enough to sing the entire alphabet song at a normal rate.

- **CLEAN SURFACES REGULARLY.**

- **USE NATURAL INGREDIENTS, SUCH AS VINEGAR, BAKING SODA, AND ESSENTIAL OILS, TO KILL BACTERIA AND VIRUSES.**

- **DON'T TOUCH MUCOUS MEMBRANES WITH DIRTY HANDS.**

- **CURB THE SPREAD OF GERMS BY COUGHING AND SNEEZING INTO THE CROOK OF YOUR ARM.**

- **WASH HANDS AFTER USING THE RESTROOM AND BEFORE COOKING.**

- **CLEAN COOKING SURFACES WHEN HANDLING RAW FOOD.**

WHAT IS CLEAN?

Cleanliness refers to hygiene, a societal concern of long standing.[81] Humans are not the only ones to be concerned about hygiene; many species exhibit activities whose sole purpose is to aid in the health of the individual and community.[82, 83] We haven't always understood hygiene, though, and throughout human history, we have struggled with the concept of and practices associated with keeping clean.[84, 85, 86] The lack of sanitation in many early civilizations led to serious health concerns.[87]

Thankfully, our knowledge of sanitation and hygiene has grown, and we now know what is needed to keep ourselves and our families healthy. As one author points out, "improvements in hygiene, cleaning, and disinfection are major contributors to the increase in life expectancy in developed countries in the twentieth century."[88]

So can you really get your home clean with just soap and water? You can clean almost everything in your house with soap and water, which is just about all anyone ever used until the twentieth century, when detergents were invented.[89] You don't need superpowered cleaners to wash away normal dirt and grime. Soap and water will wash away many of the bacteria that find their way into our homes. When you do need a little extra bacteria-killing power, there are easy, natural ways to get it that will keep you healthy.

Test an inconspicuous area of any surface you are cleaning when using a new recipe (especially on carpet or wood). But in general, the simple base recipes will be the ones you reach for time and again when you are cleaning your house.

ALL-PURPOSE CLEANERS

Do all-purpose cleaners work for everything? Almost. Certainly, if you have an area with special needs, check out some of our heavier cleaners (which are still made from all-natural ingredients). For your basic cleaning needs, however, make an all-purpose cleaner you like and start cleaning!

> Replace your current all-purpose cleaner with a 1:1 vinegar-and-water solution.

ALL-PURPOSE VINEGAR SPRAY

1 cup white distilled vinegar
1 cup water

Mix vinegar and water together in a spray bottle. Spray on desired surfaces and wipe away with a clean cloth.

CITRUS ALL-PURPOSE SPRAY

2 cups water

2 cups white distilled vinegar

15–20 drops citrus oils (orange and lemon work well)

Mix in a spray bottle and spray as needed. Note: If you use cleaning wipes frequently, try soaking cloth wipes in the solution and storing them in an airtight container.

AMAZING ALL-PURPOSE CLEANER

2 tablespoons white distilled vinegar

1 teaspoon liquid castile soap

2 tablespoons baking soda

2 cups warm water

Mix the first three ingredients, and wait for the baking soda and vinegar to stop reacting. Then add the warm water, and gently shake to mix well. Use on surfaces to clean and disinfect. You can make this up quickly any time you need it.

NATURAL DISINFECTION

Legend has it that during an outbreak of bubonic plague in the Middle Ages, robbers in the French town of Marseilles stole from the belongings of the plague's victims. The robbers, according to legend, "were spice traders and perfumers. To avoid the deadly plague, they put the magic of their essential oils (such as garlic, eucalyptus, lemon, rosemary, and sage) to work by washing themselves with the infection-fighting liquid." Their formula has survived because it is so effective.[90] Referred to as *Thieves Oil*, this all-natural blend makes a great nontoxic disinfectant. Add it to other cleaning recipes for a healthy, natural alternative to commercial disinfectants. You can purchase Thieves Oil blends or make it yourself.

THIEVES OIL BLEND

40 drops clove bud essential oil
35 drops lemon essential oil
20 drops cinnamon bark essential oil
15 drops eucalyptus essential oil
10 drops rosemary essential oil

Put all essential oils in a dark glass container. Close lid tightly and shake to mix all oils well.

THIEVES OIL BLEND DISINFECTANT SPRAY

1 cup water
8–10 drops Thieves Oil Blend

Mix water and Thieves Oil Blend together in a spray bottle. Shake well before each use. Spray on surfaces or in the air when disinfection is needed.

> For seasonal cleaning, use different essential oils at different times of the year. You may want to use lighter oils (perhaps florals) in the spring and richer oils (perhaps pine or cinnamon) in the winter.

LEMON-LIME DUST SPRAY

¼ cup lemon juice
¼ cup water
5 drops lime essential oil
5 drops lemon essential oil

Combine all ingredients in a spray bottle and shake well. Spray onto dusty surfaces and wipe away with a soft cloth. Because lemon juice is included, any extra solution must be stored in the refrigerator.

CEDAR DUST SOAP SOLUTION

If you have a particularly dusty and dirty surface, try this soap solution.

¼ cup castile soap
¾ cup water
20 drops cedar oil

Mix all ingredients in a spray bottle. Spray onto surface and wash off with a wet cloth. Shake well before each use.

Better Than Bleach

Bleach is well known for its disinfectant properties,[91, 92] as well as its harmful vapors and health effects.[93] Bleach can kill many strains of bacteria and viruses, but it can cause many health problems at the same time. Is there an alternative? Keep reading.

Dr. Susan Sumner at the Virginia Polytechnic Institute and State University showed that using a solution of white distilled vinegar followed by a hydrogen peroxide solution kills viruses better than household bleach does, without the harmful side effects associated with bleach.[94] Simply clean with a 50 percent vinegar solution as you normally would (spray and wipe or scrub); then clean the surface again with a household hydrogen peroxide (spray and wipe or scrub). This one-two punch (in either order) kills Salmonella, Shigella, E. coli, and other bacteria, along with almost all viruses and other microbes.

Note: These solutions should be used one after the other, *not* mixed together. Mixing vinegar with hydrogen peroxide results in the formation of peroxyacetic acid, also known as peracetic acid. Peroxyacetic acid is a strong oxidizing agent and can cause many of the same problems associated with bleach.[95, 96]

BETTER-THAN-BLEACH DISINFECTANT SPRAY

1:1 base recipe (All-Purpose Vinegar Spray, page 56)
Hydrogen peroxide

Spray the 1:1 base recipe (All-Purpose Vinegar Spray) on the surface you'd like to clean, and then wipe or scrub away. Spray hydrogen peroxide on the same surface, and then wipe or scrub away.

HOMEMADE DISINFECTING WIPES

Ready-to-grab cleaning wipes can make cleaning easy, but you don't need to reach for wipes with toxic chemicals. Make your own cleaning wipes at home.

1 cup water
2 tablespoons white distilled vinegar
2 tablespoons castile soap
20 drops of any of the following essential oils (mix and match!): Thieves Oil Blend (page 58), thyme, tea tree, lemon, lavender, or rosemary

Mix together and put in airtight container (a canning jar works well) with cotton wipes. Store in a cool, dark area. Shake well before using.

 Replace your Clorox wipes with homemade disinfecting wipes by putting cloth wipes and homemade disinfecting solution in a glass jar.

MIRRORS AND GLASS

Until the 1940s, our grandparents were not using much more than water and a towel to clean their glass surfaces. Today's glass cleaners contain any number of ingredients that do everything from binding to dirt (so that it comes off easier) to providing fragrance (so you don't have to smell the other chemicals). Commercial glass cleaners usually do a great job of taking away dust while leaving a streak-free shine, but at what cost?

BACK TO BASICS

The safest way to clean your glass is to get back to basics: Use water and something to wipe it off with. Use a lint-free cloth to wipe the glass down, and if needed follow that with crumpled-up newspaper to get a streak-free shine. Have you already cut out the newspaper? Tear a sheet out of that old phone book. Or use a coffee filter (preferably one made from recycled paper).

According to the Oregon Department of Environmental Quality, the average American uses 40 pounds of cleaning products every year. That adds up to more than 9 billion pounds of chemicals being dispersed into our air, released into our soil and water supply, and absorbed into our clothes, homes, and bodies.

The glass cleaner recipes are easy to make and use. If you are just switching to homemade cleaners, this is the most gratifying change you can make. Here are a few of our favorite recipes.

STARTING FRESH

Commercial glass cleaners contain detergents that can leave residue on your glass. When you start using natural cleaners, some of the streaks on your glass surfaces are likely the residue left over from your old cleaners. Adding natural soap to your initial mixture (such as in the Heavy-Duty Glass Cleaner below) when switching to a homemade glass cleaner should remove any lingering residue from commercial products. After that, switch to one of our other glass-cleaning recipes, as the soap will leave its own residue. Thankfully, castile soap is plant-based rather than petroleum-based.

HEAVY-DUTY GLASS CLEANER

¼ cup castile soap
1 quart water

Add the castile soap to the water and mix well. Spray on glass surface as needed and wipe dry.

Don't reach for Windex. Just add a cup of vodka to our All-Purpose Vinegar Spray (page 56) for a nontoxic glass cleaner that is tough on dirt and leaves a streak-free shine.

BASE VINEGAR GLASS SPRAY

½ cup white distilled vinegar
½–1 cup water

This simple spray will get your glass or mirrors sparkling clean. Mix white distilled vinegar with water. Just spray on and then, to prevent streaks, use crumpled newspaper or a lint-free cloth to polish until dry.

When you begin cleaning with vinegar, the smell might be unpleasant until you become accustomed to the absence of chemical fragrances. To help ease the transition, consider this fun twist on our base recipe.

CITRUS-FRESH BASE RECIPE

Add lemon, lime, or orange peels to a container of vinegar and let sit for at least one week. Use this vinegar to mix the Base Vinegar Glass Spray. Strain to prevent your spray bottle from clogging.

CITRUS-FRESH ALTERNATIVE GLASS SPRAY

½ cup fresh lemon juice (strain the pulp out)
1 quart Base Vinegar Glass Spray (page 65)

Add the lemon juice to your Base Vinegar Glass Spray and mix well. Spray as needed and wipe dry.

FRAGRANT EXTRACT GLASS SPRAY

10–20 drops of your favorite homemade extract (try peppermint, cinnamon, or one of the citrus extracts; see "Clean Air," page 159)
1 quart Base Vinegar Glass Spray (page 65)

Add the extract to your Base Vinegar Glass Spray and mix well.

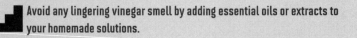 Avoid any lingering vinegar smell by adding essential oils or extracts to your homemade solutions.

UPLIFTING GLASS SPRAY

10–20 drops of your favorite uplifting essential oil (try lavender, grapefruit, bergamot, or Roman or German chamomile)
1 quart Base Vinegar Glass Spray (page 65)

Add the essential oil to your Base Vinegar Glass Spray and mix well. Shake well before each use.

STREAK-FREE GLASS AND MIRROR SPRAY #1

Adding cornstarch to your base recipe will help reduce streaks on glass.

1 tablespoon cornstarch
1 quart Base Vinegar Glass Spray (page 65)

Add the cornstarch to your Base Vinegar Glass Spray a little at a time, mixing well as you go. Heating the base spray will help the cornstarch to dissolve.

STREAK-FREE GLASS AND MIRROR SPRAY #2

Grain alcohol will help reduce streaks. The fast evaporation of the alcohol prevents streaks and spots from forming as the water dries.

1 cup grain alcohol (vodka)
1 quart Base Vinegar Glass Spray (page 65)

Add the vodka to your Base Vinegar Glass Spray and mix well.

ESSENTIAL OILS FOR DISINFECTION
Many essential oils have antimicrobial properties, meaning they can inhibit the growth of, or even kill, certain bacteria and fungi. Adding any of these essential oils to glass cleaners or other surface cleaners will remove not only fingerprints but other potentially illness-inducing germs as well.

GERM-FIGHTER GLASS SPRAY
10–20 drops of any of the following essential oils (mix and match!): cinnamon, clove, tea tree, or thyme
1 quart Base Vinegar Glass Spray (page 65)

Add the essential oils to your Base Vinegar Glass Spray and mix well. Shake well before each use.

 Cut down on the number of cleaning supplies you need. Our nontoxic all-purpose solutions and glass-cleaning solutions can be used interchangeably.

SIMPLE GLASS CLEANER ALTERNATIVE

Some people swear by club soda to clean their glass. Simply pour club soda into a spray bottle, spritz on your glass, and wipe with a lint-free cloth. Why might it work? The most plausible theory is because it contains sodium citrate, an ingredient in many detergents, which uses mineral power to increase its effectiveness.[97] We're pretty sure you could simply use water, though, and save a few bucks.

CLEANING THE KITCHEN

The kitchen is likely the most germ-ridden room in your house, regardless of how well you clean. One study found that there is more fecal bacteria on kitchen sink handles than there is anywhere in your bathroom![98] Study after study has shown that kitchen surfaces, particularly sponges and dishcloths, faucet handles, cutting boards, and the sink, are the areas where bacteria proliferate and spread easily from person to person.[99]

When you start switching to natural cleaners in your kitchen, the easiest change to make is simply to use one of our all-purpose cleaners on your sinks, countertops, and appliances. All-Purpose Vinegar Spray (page 56), Citrus All-Purpose Spray (page 57), and Amazing All-Purpose Cleaner (page 57) will all be effective, but the Thieves Oil Blend Disinfectant Spray (page 58) is particularly well suited to the kitchen.

If you're looking for more specific recipes to clean specific surfaces in your kitchen, including wood, stone, or stainless steel, keep reading.

CUTTING BOARDS: FOOD PREP

Use cutting boards to help keep bacteria localized.

A source of heavy contamination in the kitchen is the cutting board. Raw meat has a much higher concentration of bacteria than vegetables, fruit, or bread, and special care should be taken when preparing and cooking meat. When preparing different types of food, the safest option is to have multiple cutting boards: one for meat, one for produce, and a third for bread, if you so desire. If your kitchen has butcher block counters for food preparation, consider having at least one removable cutting board exclusively for raw meat that you can easily wash in the sink.

Research is split on whether wood or plastic cutting boards are safer. Bacteria can get lodged into tiny cuts and nicks in plastic and into the grain of wood, which is porous. Although wood cutting boards may retain more bacteria than plastic, which is nonporous, plastic boards may transfer bacteria more readily than do wood boards.[100] Less research exists on other nonporous cutting board materials, such as glass or pyroceramic, possibly because those materials are hard on knives and so are not as popular as plastic and wood boards.

Different types of cutting boards have different benefits. Wood is naturally antibacterial and biodegradable, and it takes fewer resources and energy to make than plastic. BPA-free plastic cutting boards are often thinner, making it easier to have multiple cutting boards in a small kitchen and prevent the spread of bacteria. Regardless of the type of cutting boards you choose, it is important to properly wash them to kill any lingering bacteria.

CLEANING PLASTIC AND OTHER NONPOROUS CUTTING BOARDS

Plastic and glass cutting boards can be washed in the dishwasher, but research shows that bacteria from the cutting board can contaminate other dishes in the dishwasher. It might be safer to clean plastic the same as wood: scrub with very hot, soapy water.[101] For extra protection, follow this wash by spraying your cutting board with vinegar and wiping it with a clean cloth.

CLEANING WOOD CUTTING BOARDS AND BUTCHER BLOCK COUNTERS

To clean a wood cutting board, put it in the sink with very hot, soapy water. Scrub with a bristle brush, and then rinse it and allow it to dry. To clean butcher block countertops, scrub with hot, soapy water and a brush or sponge. For light cleaning of butcher block, spray with vinegar and wipe with a clean cloth. Whether you use water or vinegar, thoroughly dry your cutting board and butcher block after cleaning them. Wet wood encourages bacterial growth and can cause warping or other damage. Do not put wood cutting boards in the dishwasher, and do not use wire brushes on wood cutting boards or butcher block. Either action can damage the wood.

SALT AND LEMON BUTCHER BLOCK AND WOOD CUTTING BOARD CLEANER*

Coarse salt
Lemon juice

For a deep clean, sprinkle coarse salt liberally on the board. Squeeze the juice from one lemon onto it, and then scrub for several minutes with a bristle brush. Rinse and dry.

Butcher blocks are great to cut on but can harbor bacteria. The acid in many citrus fruits will kill a lot of the bacteria that tends to grow on butcher block and inhibit future growth. Try scrubbing with lime or orange juice instead of lemon juice for a different citrus smell.

BAKING SODA AND VINEGAR BUTCHER BLOCK AND WOOD CUTTING BOARD CLEANER*

Baking soda
Distilled white vinegar

Instead of salt and lemon, sprinkle baking soda liberally on the board and add a couple of tablespoons of vinegar. Then scrub for several minutes with a bristle brush before rinsing and drying.

* You could also use salt and vinegar or baking soda and lemon. The salt and baking soda are abrasives that will help loosen grime and lift it away; the vinegar and lemon have antibacterial properties to help kill bacteria and combat odors.

COUNTERS

Counters, whether in the bathroom or kitchen, are places where bacteria like to linger. To minimize bacterial growth, clean all kitchen surfaces immediately after preparing food. Just about every kitchen surface can be effectively cleaned by a good scrub with soap and hot water. For solid surface counters (those made from acrylic and other plastics), try one of the all-purpose recipes (page 56). Keep reading for more recipe ideas for some of the most popular (nonplastic) counter materials.

WOOD

Unlike some other kitchen counters, wood butcher block counters can also be used as a cutting surface. Naturally antibacterial, the wood still needs to be thoroughly cleaned and should be treated with natural oils, rather than commercial petroleum- or polyurethane-based products. Consumers have traditionally used mineral oil to treat butcher block, but it is derived from petroleum; hence it is not easily biodegradable, nor is it a sustainable choice.[102] The same is true for paraffin wax.[103] Instead of mineral oil, try all-natural oils to condition your wood.

Robin Wood, an internationally respected woodworker with twenty years of woodworking experience, describes good oils as being as variable as good wines. Wood uses only oils that are grown near him and that are "safe enough to drink." He uses linseed and walnut oils exclusively on his own creations, but he is also comfortable with tung oil as a natural substitute for mineral oil. Tung oil is derived from flaxseed. All three are natural, are food-safe, and generally will not go rancid (or at least they will not go rancid as easily as many plant-based oils can). Because these are drying, or "curing," oils, they will actually harden in the wood over time and create a thin, water-resistant coating on the wood. Reapplication is needed periodically.[104]

Tung oil may be harder to apply evenly, so use care.[105] Linseed oil's coloring tends to give wood a yellow hue over time. If you purchase linseed oil, do not buy boiled linseed oil, which contains chemical driers. To avoid chemical additives in any of these three varieties, purchase raw, pure oils. You can find walnut oil in your grocery store's oils section. Tung and linseed oils can be purchased online or in your local pharmacy or craft store (again, be sure you are purchasing pure oils without additives).

Caution: Drying oils left on rags generate heat as the rags dry, and they can easily spontaneously ignite. Linseed oil–soaked rags have been the cause of many fires. Oil-soaked rags have even been known to ignite in the dryer after being washed. Follow proper disposal directions as listed on the oil container, or contact your local hazardous household waste disposal program.

> Eliminate toxins on your counters: Treat butcher block counters with pure coconut or walnut oil.

Coconut oil is another food-safe oil many individuals use on counters. It is very stable and unlikely to go rancid, as can happen with some less stable food-based oils. The nice thing about wood is that if you use something you don't like, you can simply sand it off and start fresh with something else.

Oil your butcher block counters once a month (if they are new, oil them daily for a week, and then oil them weekly for a month, and then move to monthly), after thoroughly washing the counter to kill bacteria. To use your natural oil, place the bottle of oil in hot water for several minutes to warm the oil. Pour a good amount on the wood, and distribute the oil evenly with a soft cloth. Let the oil sit overnight (about 8–10 hours), and then wipe off any excess with a clean, dry cloth. If using drying oils, do not let the rags sit in your house. Follow directions for proper disposal of the oil-soaked rags.

STONE

Natural stone counters are a popular choice for kitchens. Granite, marble, quartz, slate, and travertine are just a few of the many possibilities. There are several universal rules for natural stone counters:

- WIPE UP SPILLS IMMEDIATELY.

- DO NOT USE ABRASIVE OR ACIDIC CLEANERS.

- DO NOT SET HOT OR WET OBJECTS DIRECTLY ON THE STONE (USE A POT HOLDER, TRIVET, OR SOMETHING SIMILAR).

Engineered stone counters, most often manufactured from natural quartz and a binder, are often more scratch- and heat-resistant than natural stone. It is still a good idea, however, to treat any stone counter with care.

Keep a bottle of Heavy-Duty Glass Cleaner (page 64) on hand for light cleaning of both natural and engineered stone counters. Spray your counter and wipe with a clean cloth. For heavier cleaning, use soap and very hot water, or try one of the recipes that follow. Use a soft cloth or a brush with soft bristles; anything more abrasive can scratch the stone.

Coconut Oil Counter Rub

Some porous counters, including granite and butcher block, do best when they are treated on a weekly basis. You don't need to purchase products laden with harmful products, however. An easy way to treat these surfaces is to rub them with coconut oil. Coconut oil is naturally antimicrobial,[106, 107] antibacterial, [108] antiviral, and antifungal.[109]

Coconut oil has a low melting point, so depending on the temperature of your home — which often varies depending on the time of year — your oil will be either a liquid or a solid. Whether liquid or solid, you need just a small amount to rub into the counter. After allowing it to soak in, you may notice lines appearing where it has absorbed into the pores. Use a clean, dry cloth to buff it. Your counter will be treated and resistant to all of the microbes invading the surface.

GRANITE COUNTER CLEANER

⅛ cup vodka
1 cup water
Drop of dish soap
A few drops of essential oil of choice

Mix ingredients in a small spray bottle. Shake well before use.

Marble

Marble, a natural substance formed when limestone is exposed to heat or pressure, is classified as a soft rock and is largely made up of calcium carbonate.[110] Although marble looks hard, it can be easily damaged by acidic substances. Never use vinegar or lemon juice to clean marble! Instead, use soap and water for ordinary cleaning. To remove organic stains (coffee, wine, and so on) from marble, try the following recipe, a trick that the United States government uses to clean historical marble statues.

MARBLE STAIN REMOVER

Hydrogen peroxide
A white absorbent material: crushed white chalk, unbleached white flour, or white paper towels

Clean the area with water. Wet the stain with hydrogen peroxide. Next, mix enough hydrogen peroxide with the absorbent material to create a poultice, or paste. Apply the poultice in a thick layer on and around the stain. Cover the poultice with a piece of plastic, and use tape to secure it. Let the poultice sit for 48 hours, and then moisten it with water and remove it with a plastic spatula (metal may scratch your marble).[111]

METAL

Metal countertops, including stainless steel, zinc, pewter, copper, and bronze, are beautiful options for kitchens. They are durable and do not require seals or finishes. Some metals are heat- and stain-resistant (particularly stainless steel), some develop a beautiful patina if you do not polish them regularly, and others are recyclable.

Copper and copper-containing alloys are an especially healthy choice because copper is highly antibacterial. Research has demonstrated that all E. coli bacteria placed on a copper surface died within 90 minutes (refrigerated) or 270 minutes (at room temperature). Copper-containing alloys took 120 minutes (refrigerated) and 360 minutes (at room temperature) to kill the bacteria. It took 28 days for the E. coli to die on stainless steel at either temperature.[112] Similar results were found when researchers measured the survival of a strain of influenza virus. Researchers placed 2 million particles of the influenza A virus on stainless steel and copper. After 6 hours, only 500 particles were still alive on copper; 500,000 particles were still alive on stainless steel after 24 hours.[113]

STAINLESS STEEL COUNTER CLEANER

1 part cream of tartar
1 part hydrogen peroxide

Mix cream of tartar and hydrogen peroxide into a paste. Apply to counter and wipe off with a damp cloth. Buff with a clean, dry cloth for extra shine.

COPPER COUNTER CLEANER

½ lemon
Coarse salt

Cut a lemon in half and sprinkle it with coarse salt. Rub over the copper counter, and then wipe with a damp cloth.

OTHER COMMON COUNTER CHOICES

There are many other choices for kitchen counters. Ceramic tile can add pops of style and color, but be wary of imported tile, which may contain lead. You must seal tile and grout periodically to resist mildew and stains. Like stone, ceramic tile and grout may be damaged by acid, so avoid using vinegar- or lemon-based cleaners on them.

Porcelain and ceramic are traditional materials for bathroom counters. Durable and easy to clean, you can generally use one of the non-acidic scrub or disinfecting recipes from the Bathroom chapter starting on page 101.

TILE, PORCELAIN, AND CERAMIC COUNTER CLEANER

Heavy-Duty Glass Cleaner (page 64)
10–20 total drops essential oils (choose two or more from among Thieves Oil Blend (page 58), thyme oil, and tea tree oil)

Pour the Heavy-Duty Glass Cleaner in a spray bottle, add the essential oil, and shake well.

BASIC CLEANER BOOST

Thieves Oil Blend, thyme oil, or tea tree oil can all give the basic cleaner an added antibacterial boost for the areas of your home where you need it most.

SINKS

Any area that is exposed to moisture, particularly sinks and the tools you use to clean dishes, are prone to bacterial growth and transfer. Slow the growth of bacteria by keeping these areas dry. Keep a basket of T-shirt towels under your sink, and wipe the sink and counters down with a clean, dry cloth after doing the dishes. Then toss the towel and the dishcloth into the laundry.

Every few days, take out your sink's drain catcher (the removable rubber or metal part) and give it a good rinse and scrub. Keep it clean by boiling it along with your other sink tools (see "Washing Dishes," page 85). While you're waiting for your sink tools to boil, use one of the recipes that follow, and give your sink a good scrub. Your sink will sparkle, you'll cut down on bacterial growth, and your kitchen will smell divine.

SINK ESSENTIALS

The Essential Sink Cleaner works wonderfully for stainless steel and porcelain sinks. Use soft cloths, sponges, or soft bristle brushes to clean stainless steel sinks to avoid harming the finish.

ESSENTIAL SINK CLEANER

½ cup white distilled vinegar
¼ cup baking soda
5 drops of your favorite essential oil

After cleaning the sink of any debris and wiping it down with water, wet the sink with the vinegar. Sprinkle on the baking soda and a few drops of your favorite essential oil. As you scrub the sink with your cloth, the baking soda and vinegar will form a paste. Afterward, be certain to rinse well to avoid leaving a residue.

Easily sanitize the dirtiest part of your house — the kitchen sink — with our Sweet Cinnamon Sink Scrub. Together, baking soda, cinnamon, and sweet orange essential oil kill bacteria and germs.

SWEET CINNAMON SINK SCRUB

This scrub smells wonderful and is perfect for fall and winter, though you can use it anytime. Use the oils or other ingredients that smell good to you. When cleaning is pleasant, you are more likely to keep up with tasks.

1 cup baking soda
1 tablespoon ground cinnamon
5 drops sweet orange essential oil

Mix all ingredients in an airtight container. To use, sprinkle some of the Sweet Cinnamon Sink Scrub on a wet sink. Use a cloth to scrub the sink. Rinse well.

SINK STAIN SCRUBBER

½ cup borax
½ cup baking soda
10–15 drops tea tree essential oil
Vinegar

Carefully mix the dry ingredients and essential oil in an airtight container. Sprinkle on sink stains and scrub with a damp cloth. Use the vinegar to rinse. Follow up with a hot water rinse.

SUPER LEMON STAIN REMOVER

½ cup baking soda
10 drops lemon oil
Lemon juice

Make a paste using all ingredients. Place on the stain and allow the mixture to sit for a few hours or overnight. Gently scrub the spot and rinse well.

CLOGGED SINK CLEANER

When your drains are clogged, skip pungent chemicals that can cause illness and damage your pipes. Baking soda and vinegar will help unclog the drain, and the boiling water will finish loosening and flushing the debris away. To keep drains free of hair and other particles, purchase a relatively inexpensive snake from your local hardware store. Use it to clean drains semi-annually or as needed.

½ cup baking soda
1 cup vinegar
1 gallon boiling water

Pour baking soda into the drain, add vinegar, and allow to bubble and react for 15 to 20 minutes. In the meantime, boil at least 1 gallon of water, then flush the drain with the boiling water. Repeat as needed.

GARBAGE DISPOSALS

- **REMEMBER TO RUN COLD WATER WHILE USING YOUR DISPOSAL.** You want any grease that goes down the drain to solidify and get chopped up by the disposal. Hot water will cause grease to liquefy; when this occurs, grease might pool in your pipes, causing clogs and bad odors.

- **RUN YOUR GARBAGE DISPOSAL REGULARLY.** Letting it go too long before running it can allow rust to accumulate, rendering your disposal useless.

- **TAKE IT EASY.** The disposal is for your convenience to deal with those food scraps that make it into the sink. Nonfood items, large items, and large quantities of food should be disposed of properly: in the compost or in the garbage.

- **LET THE WATER RUN.** If your garbage disposal is running, the water should also be running. Let the water continue to run for at least 15 seconds after turning off the disposal in order to prevent a buildup of small particles. Not only can these potentially clog your disposal, rotting food can cause a stench you don't want in your kitchen.

ICY COLD DISPOSAL CLEANER #1

Handful of ice cubes
½–1 cup regular or rock salt

Toss the ice cubes and salt into your garbage disposal. Let sit for a few minutes, turn on the cold water, and run the disposal until all the ice breaks down and finally disappears. The ice chips will scour your disposal, cleaning away grime.

ICY COLD DISPOSAL CLEANER #2

Freeze vinegar in ice cube trays. Put a handful of vinegar ice cubes into the drain, along with 1 cup of baking soda. Run cold water and turn the disposal on until all the ice breaks down and finally disappears. The ice chips will scour your disposal, and the baking soda and vinegar combo will help clean away odors.

NATURAL DISPOSAL SCOURER

Use eggshells to naturally scour your garbage disposal. The particles created when the eggshells are broken apart by the disposal will act as a gentle abrasive, cleaning the sides of the disposal.

CITRUS PEEL DISPOSAL CLEANER

Take 2 or 3 oranges, lemons, or limes that are about to go bad. Cut them in half and feed them, one at a time, into your disposal with the cold water running. Turn the disposal on and let it chop each half, rind and all. The rinds will help scour away grime while the citrus oils will help combat odors.

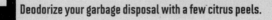

Deodorize your garbage disposal with a few citrus peels.

BAKING SODA DISPOSAL DEODORIZER

Pour 1 cup of baking soda into the disposal. Let it sit for at least 1 hour. Pour 1–2 cups of vinegar into the disposal, and run the disposal while the vinegar bubbles.

BORAX DISPOSAL DEODORIZER

Pour 4 tablespoons of borax into the disposal. Let it sit for 1 hour. Follow with hot water, vinegar, or both.

EASY DISPOSAL WASH

Washing your garbage disposal couldn't be easier. While running the cold water and the garbage disposal, squirt some liquid dish soap down the drain. In the amount of time it takes to wash your hands, your disposal will be clean.

WASHING DISHES

Much of the bacteria in your kitchen comes from food. The bacteria is spread as you prepare and cut food and then clean up your mess. Bacteria can travel from the raw meat, to the cutting board, to the sponge, to the sink handle, to the hands of the next person who uses the sink, to whomever or whatever that person touches next. To cut down on the spread of bacteria, take care to clean your food prep surfaces and utensils immediately and thoroughly with soap and hot water after they are used.

You don't need a fancy dish soap to wash dishes. You can use a 1:10 ratio of castile soap and water. Simply put it in a squirt bottle, and use it on your sponge or scrub brush. You'll probably want to use a vinegar water rinse on your dishes, or you may notice a thin film on them. The film is actually from salts that are left after the castile soap reacts with minerals in the water, which is especially problematic if you have hard water. Vinegar will help cut those salts. After rinsing the soapy dishes off with water, have another rinse bucket with a quart of water and a cup of vinegar ready. Give the dishes a good dunk and swish before placing them on the drying rack. Instead of dunking them, you can also try spraying them with a vinegar-and-water mixture on the drying rack. Unfortunately, it does not work to add vinegar (an acid) directly to your soap (a base) and water, as the vinegar will react with the soap.[114]

If you don't want to use just castile soap alone, try one of the following recipes for dish soap.

ESSENTIAL DISH SOAP

¼ cup tightly packed, grated bar soap
1¼ cups boiling water
1 tablespoon washing soda
¼ cup castile soap
20–30 total drops essential oils of your choice

Add the grated soap to the boiling water. Stir until dissolved. Add washing soda and castile soap. Stir well and remove from heat. Allow the mixture to cool, and add the essential oils. Store in a glass jar or soap dispenser.

SIMPLE CITRUS DISH SOAP

Skip store-bought dish soap and try the Simple Citrus Dish Soap instead. Squirt some in to your next sink full of dirty dishes.

20 ounces castile soap
20 drops lime or lemon essential oil
20 drops sweet orange essential oil

Mix all ingredients together. Squirt bottles work exceptionally well for dish soap. Add a squirt or two, as needed, the next time you wash dishes.

FLORAL DISH SOAP

20 ounces castile soap
10 drops lavender oil
10 drops ylang ylang oil
10 drops rosemary oil

Mix all ingredients together and use a squirt or two, as needed, when washing dishes.

DISH WASHING TOOLS

Sponges, with their deep holes that trap food particles and their tendency to stay damp, are ideal breeding grounds for bacteria.[115] Additionally, most commercial sponges are made from petroleum derivatives and are not good for the environment. Following are a few alternatives, all of which are available through several online retailers, that can help you cut down on bacterial growth.

> Boil kitchen sink tools to eliminate bacteria, such as E. coli and salmonella. Drop sponges, sink stoppers, and scrub brushes in boiling water for up to 5 minutes, or put them in the dishwasher.

Hemp Scrubbers

Hemp dishcloths are generally more abrasive than cotton but soft enough for regular dishwashing needs. Hang them to dry between uses, and throw them in the laundry every day or two.

Cotton Dishcloths

Buy dishcloths that have some texture to them, or knit your own. Hang to dry after use, and launder every day or two.

Coconut Scrubbers

There are several types of scrubbers and pads made from coconut coir fiber, extracted from the husks of coconuts. They are great cleaners and the coconut fiber is biodegradable. As with any sponge or scrubber, disinfect them regularly.

Loofah Scrubbers

Loofahs are made from gourds, so they are biodegradable. They will still trap bacteria, so be sure to dry them after use, sterilize them at least once a week, and compost them when you're done with them.

Sponges

If you are attached to sponges, opt for biodegradable cellulose sponges, and replace them as soon as they start showing signs of wear. In one study, researchers found heavy concentrations of bacteria on every kitchen surface examined in a particular home, day after day, "until the sixth day, when most surfaces suddenly turned up virtually germfree. It turned out the family had simply begun using a new sponge."[116]

Keep dishcloths, sponges, and scrub brushes in an area (such as a hanging basket) where they can dry between uses, preferably in the sunshine. To sterilize them, throw them all into a pot of boiling water for up to 5 minutes and then hang to dry.

Alternatively, put your sponge in your dishwasher for a full wash and dry cycle, or wash your dishcloths in the washing machine (add vinegar or tea tree oil to the rinse cycle). Research is split on whether microwaving a wet sponge for 1 minute at full power is the best way to kill bacteria, so you may want to stick with boiling as a sure way to kill any bacteria on your sponge. Better yet, put all of your kitchen supplies that need sterilized in one pot of boiling water. You'll get them all clean at once, and you won't have to worry about the possibility of setting your sponge on fire in the microwave.[117]

DISHWASHERS

When done correctly, washing dishes in an automatic dishwasher is more efficient than washing them by hand. According to dishwasher manufacturers, this is how to wash dishes in the dishwasher:

- **DO NOT RINSE DISHES.** Simply scrape food into a trash can or compost pail.

- **WASH FULL LOADS OF DISHES TO OPTIMIZE WATER AND ENERGY USAGE.**

- **USE ANY ENERGY-SAVING FEATURES AVAILABLE ON YOUR AUTOMATIC DISHWASHER.** Having an energy-efficient hot water heater is also beneficial.

You can consume less energy and water by washing dishes manually than by using some automatic dishwashers if you do more dishes at one time and if you limit your water use. (Wash your dishes in one bowl, and rinse them in a second bowl.)[118]

If you do own an automatic dishwasher, following are several recipes you can use to keep your dishes and dishwasher clean:

SIMPLE DISHWASHER DETERGENT

1 cup borax
1 cup washing soda
½ cup kosher salt
½ cup citric acid

Mix all ingredients well, being careful not to inhale as you mix. Use 1 tablespoon of detergent per load. Store in an airtight container.

LEMON-POWERED DISHWASHER DETERGENT

2 cups borax
2 cups baking soda
1 cup citric acid
20–30 drops lemon essential oil

Carefully mix all ingredients together and store in an airtight container. Be careful not to inhale the fine powder as you mix. Use 1 tablespoon per load.

LAVENDER DISHWASHER DETERGENT

2 cups washing soda
1 cup borax
1 cup baking soda
20–30 drops lavender oil

Carefully mix all ingredients together, making certain not to inhale the powder, and store in an airtight container. Use 1–2 tablespoons per load.

DISHWASHER GEL

2 cups water
¼ cup soap flakes or grated bar soap
1 tablespoon glycerin
2 tablespoons white distilled vinegar

Put all ingredients in a saucepan and bring to a boil over medium heat. After all the soap has dissolved, remove from heat and allow to cool completely. Store in an airtight container and use 1 tablespoon per load.

LEMON DISHWASHER GEL

2 cups castile soap
½ cup water
½ cup lemon juice
20 drops lemon essential oil

Mix all ingredients together in a saucepan over medium-high heat, stirring until all ingredients are dissolved. After cooling, store in an airtight container. Use 1 tablespoon per load.

SIMPLE SPOT REMOVER

Your dishes may be clean, but if spots are plaguing the glasses washed in your dishwasher, you may find yourself considering a commercial spot remover. No need! Add a diluted vinegar solution to the rinse aid compartment; the rinse should take care of the spots.

3 parts white distilled vinegar
1 part water

Dilute vinegar with water and place in rinse aid compartment. Run dishwasher as normal.

Some individuals like to use straight vinegar in their dishwasher as a rinse aid. Although 100 percent vinegar will get your dishes spot-free, vinegar's high acidic content might damage the rubber and plastic in the rinse aid compartment. Dilute the vinegar to avoid this risk.

VINEGAR DISHWASHER DEEP CLEANER

If you have a dishwasher, you probably know that over time it can need a deep clean. Food particles collect. It may start to smell. It's not pretty. Is there anything you can do? Yes! Fill two mugs with white distilled vinegar, and place them in the dishwasher, one on the top rack and one on the bottom, with the mugs sitting upright. Run your dishwasher as normal. As water pours into the mugs, the vinegar will slowly clean your dishwasher.

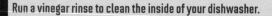 Run a vinegar rinse to clean the inside of your dishwasher.

LEMONADE DISHWASHER DEEP CLEANER

Pick up a few packets of lemon-flavored drink mix (such as Kool-Aid), and add one to your dishwasher. Run the cycle as normal. The citric acid in the drink mix will help clean your dishwasher. If you already have citric acid on hand, try using some in place of the drink mix.

POTS AND PANS

Replace nonstick cookware with copper, stainless steel, or cast iron to avoid the toxins released by Teflon coatings.

Some of the most popular cookware sets on the market, the ones with a nonstick (Teflon) coating, are hazardous to your family's health. The Teflon coating reduces burning while cooking and makes cleaning easy. But when that coating reaches certain temperatures, it releases toxic, carcinogenic fumes that have been shown to affect your health, your pets' health, and the health of your unborn baby.[119] To avoid these problems, consider some of the alternatives to nonstick cookware.[120]

Cast Iron

Cast iron is a wonderful medium to cook with. It is economical to buy and use, and it absorbs and distributes heat efficiently. Cast iron is sturdy and will last for decades. Unseasoned cast iron is porous, with tiny holes throughout the surface. Prior to use, cast iron cookware should be seasoned in order to prevent your food from burning and sticking and your pots from rusting.

Before seasoning a new piece of cast iron, clean it with soap, hot water, and a stiff (not wire) scrub brush. Dry it thoroughly. Use a soft cloth or paper towel to spread lard or vegetable oil all over the inside surface. Bake the pot upside down on a cookie sheet for 1 hour at 350 degrees Fahrenheit. Due to the oil on the pot, you may notice smoke in the air; turn on your exhaust fan and open a window.

Each time you use your cast iron, it will become more seasoned. Eventually, you will see a shiny black patina.

CLEANING CAST IRON: After using your cast iron, scrape excess food off with a spoon or a scrub brush (not a wire scrub brush), first without water and then with hot water running over the pan. If food is stuck on, pour boiling water into the pan. Let it sit and soften the food for a few minutes. Pour it out and scrape again.

Another trick you can try for stuck-on food is to scrub the cast iron with coarse table salt and a soft cloth before rinsing it with water. If you use soap and water, you will need to reseason the cast iron. Water and scraping alone should be enough to clean your cast iron.

Place the pan on a hot stove burner or in the oven for a few minutes to thoroughly dry it. Once it is dry and cool, put another thin layer of oil or lard on it, removing any excess with a soft cloth or paper towel.

Do not put cast iron in an automatic dishwasher, as this can remove the seasoned coating and cause rust to form.

STORING CAST IRON: Store cast iron in a dry place. Leave a soft cloth or paper towel in between pans to absorb excess moisture.[121]

ENAMELED CAST IRON: Enameled cast iron is cast iron covered with a thin layer of enamel. It is food-safe, durable, and does not need to be seasoned. Because you do not season enameled cast iron, you can clean these pots and pans with soap and water.

Copper

Many chefs love copper because it is a great heat conductor. It is a highly reactive metal, however, so it can react with certain foods (especially tomatoes and other acidic foods), making anyone who ingests the food ill.[122] To prevent this, look for copper cookware that is lined with stainless steel.

Wash your copper cookware with soap and water. If the copper becomes stained, try our Copper Counter Cleaner (page 78) or create a paste out of lemon juice and baking soda. Rub the stained copper with the paste and a soft cloth, and then wash off with soap and water.

Stainless Steel and Anodized Aluminum

Other cooks prefer stainless steel and anodized aluminum because they are durable, do not react with acidic or alkaline foods, and are more affordable than copper or cast iron.

To make it heat more evenly, stainless steel cookware contains other metals (often copper or aluminum). Preheating your stainless steel pot and adding a very small amount of cooking oil before cooking food often helps prevent food from sticking to stainless steel.

Anodized aluminum is simply aluminum that has been processed so that it has a nonreactive coating. Because it is nonporous, you should not need to add oil to prevent sticking. Preheating your anodized aluminum pan over medium heat before adding food may help you cook your food evenly. Note: Because research has shown certain links between aluminum and Alzheimer's disease, we recommend replacing any anodized aluminum cookware that becomes scratched or that allows food to come in contact with an aluminum surface (including edges).[123]

Clean both stainless steel and anodized aluminum cookware with soap and water. If food has burned onto your cookware and will not come off with normal washing, boil water in the pan for 10–20 minutes to loosen the food, let the pan cool, and then wash with soap and water again. Or put a tablespoon or two of baking soda into the pan, add enough water to create a paste, and then scrub the pan with a sponge or soft brush.

As a general rule for all types of pots and pans, avoid pouring cold water into a hot pan, as this can cause many pans to warp. Allow pans to cool before washing to protect yourself and your pans.

VEGETABLE AND FRUIT WASH

Most fruits and vegetables sold today are covered in pesticides. That is definitely something you don't want to be eating. Try these basic, inexpensive washes to remove pesticides from the surfaces of your food.

THE DIRTY DOZEN PLUS AND THE CLEAN FIFTEEN

Want to avoid chemicals in your food? Organic options are a great way to go, but they cost more. Stretch your budget by choosing wisely. Look online to find the "Dirty Dozen Plus" and the "Clean Fifteen" lists, released by the Environmental Working Group (EWG). The lists change each year, so check to be up to date. The Dirty Dozen Plus are the top twelve (or more) fruits and vegetables that are most affected by pesticides. These are the items you should be most concerned about. The Clean Fifteen are the fruits and vegetables least likely to be affected by pesticides, so you can save money by purchasing the conventionally (non-organically) grown versions.

VINEGAR VEGETABLE SPRAY

For hard-skinned vegetables and fruits, you can get away with spraying the food, letting it sit for a few seconds, and then gently scrubbing it clean in fresh water. Just use the All-Purpose Vinegar Spray (page 56) to help cut through the pesticides.

Soak or gently scrub fruits and vegetables with All-Purpose Vinegar Spray (page 56), and rinse with water to cut through pesticides used on conventional produce.

VINEGAR FRUIT SOAK

For softer vegetables, soak rather than scrub. Mix up your 1:1 vinegar-and-water solution, and allow the items to soak for a minute or two. Then rinse them off with fresh water.

REFRIGERATOR

Food spills are easier to clean when taken care of right away, but you don't always have time to do this or even realize the spills have occurred when there are other people in your home. Try these simple tips for keeping a clean refrigerator.

Nothing funks up a fridge as quickly as rotting, moldy foods. Do your fridge and yourself a favor by removing them in a timely fashion. There are plenty of other science experiments you can do. Botulism shouldn't be one.

LEMON FRIDGE SPRITZ

Find yourself cleaning out the gunk in the refrigerator? Try mixing up a 1:1 solution of water and lemon juice. The acid in the lemon juice will kill bacteria while leaving a refreshing scent.

Don't forget the gaskets around the refrigerator and freezer doors. These are prime locations for dirt, grime, and bacteria to build up. Use the lemon spritz and an old toothbrush to clean out these hard-to-reach areas.

CITRUS FRIDGE DEODORIZER

1 small box of baking soda
20–30 total drops of citrus essential oil (try lemon, orange, lime, or grapefruit)

Mix the baking soda and the essential oil. Place the mixture in the refrigerator in an open container to absorb any smells.

> **Clean up spills right away to make cleaning easier and avoid bacterial contamination.**

MICROWAVE

> **Microwave a water-and-vinegar mixture to loosen stuck-on food inside your microwave.**

Microwaves can almost clean themselves. Almost. In just a few steps, your microwave can be sparkling clean without the use of harsh chemicals.

QUICK-AND-EASY MICROWAVE CLEANER

1 part white distilled vinegar
1 part water

Combine the vinegar and water in a microwave-safe bowl. Microwave on high for a few minutes. The solution will steam and soften any food splatters inside the microwave. After carefully removing the bowl (both the bowl and the liquid will be very hot), use a soft cloth and your favorite all-purpose spray to clean out the microwave. Everything should wipe right out.

Clean the outside of the microwave with one of your all-purpose sprays (page 56). If your microwave is located above your stove, don't forget to periodically take off any vents or filters and clean those. Because the vents are pulling air through that has grease particles in it, your best bet is to use hot, soapy water to clean away the grease. Doing so will help the ventilation system work properly and limit the amount of grease that spreads around your kitchen, resulting in less cleaning for you.

STOVE AND OVEN

> Make cleaning burned-on food from a stove top easy by pouring hot water on it, letting the water sit for several minutes, and wiping the stove top clean with a towel.

The heat used for cooking can you leave your stove or oven with a baked-on mess. You don't need to resort to toxic chemicals or build up your muscles to clean it, though. There are some simple and effective ways to get it clean.

GLASS STOVE CLEANER

1 part baking soda
1 part fine salt
1 part water

Mix all ingredients together to form a paste. Put the paste on the glass stove top and allow to sit. Gently scrub with a soft cloth.

HOT WATER STOVE CLEANER

This sounds incredibly simple—and it is! Nothing cleans off baked-on, hardened food from your stove top better than hot water. Simply pour hot water on the stove top and allow it to soften the food. After wiping with a soft cloth, do a once-over with one of your all-purpose sprays to kill any remaining bacteria. If your water is hot enough, you can even skip this part.

OVEN CLEANER

Warm or hot water
Baking soda
Orange essential oil

Add warm or hot water to baking soda until you get a loose paste. Add a few drops of essential oil for scent. Spread mixture on dirty areas of the oven, and allow the paste to sit for 15–30 minutes. Using hot water and a scrub brush, clean oven thoroughly.

 Instead of toxic oven cleaners, use a hot water and baking soda paste in your oven. Let it sit, and then gently scrub off baked-on gunk with a brush.

BATHROOMS

In the bathroom, most surfaces can be cleaned with all-purpose cleaners, and counters can be given the same treatment as kitchen counters. But toilets, tubs, and a moist environment present a unique set of challenges. Don't despair. Those issues can be overcome with these simple answers to your bathroom needs.

TOILETS AND TUBS

Even though the kitchen harbors the most bacteria in the house, that's not to say your bathroom is pristine. Every time you flush your toilet, an aerosol spray of water droplets, laden with bits of feces and urine, explodes into the bathroom. The bacteria contained in those bits of waste travel as far as eight feet from the toilet, onto your bathroom sink, the floor, and even your toothbrushes.[124] Closing the toilet lid before flushing can limit the spray, but it won't completely eliminate the spread of bacteria. Nothing is safe.

Modern cleaners often rely on bleach to kill bacteria. But bleach has its own issues. Instead, try some more natural alternatives for a sparkling clean bathroom.

The bathtub cleaners in this section are designed with porcelain and ceramic in mind. If you have stone in your bathroom, such as travertine tiles in the shower, do not use anything acidic on it. Doing so can permanently damage the stone surface. Look for the recipes in this section that are notated as "safe for stone." Also, be aware that baking soda is a mild abrasive; when you scrub stone with baking soda, be gentle.

HEAVY-DUTY SOAP SCUM REMOVER

½ cup baking soda
Castile soap (enough to make a paste)
5 total drops tea tree essential oil, clove essential oil, or a combination

Form a paste with the baking soda and soap. Add essential oil. Apply to any areas of your shower (glass doors and such) that have soap scum buildup. Spray with vinegar or a 1:1 vinegar-and-water mix, and wipe away with a sponge or cloth.

* To make this recipe safe for stone, spray with water only—no vinegar.

> For an easy, nontoxic soap scum remover, make a paste from baking soda, castile soap, and tea tree or clove oil. Spray with vinegar and wipe away.

TOILET RUST REMOVER

Does your toilet have rust stains in it? Try dissolving some borax in the toilet bowl and allowing it to sit overnight. The next morning, use your toilet brush to scrub away the stains.

 Use soap, baking soda, and essential oil to safely remove rings around the tub.

RUB-A-TUB SCRUB

1 teaspoon liquid soap
1 cup baking soda
5 drops essential oil (try tea tree, eucalyptus, peppermint, or lavender for their antibacterial properties)

Mix all ingredients and add just enough water to form a paste. Using a cloth wipe or small brush, scrub the bathtub with the paste. Rinse well after cleaning to avoid any residue.

* Safe for stone

TUB AND TILE SCRUB

¾ cup baking soda
¼ cup castile soap
1 tablespoon water

Mix all ingredients together until dissolved and a paste has formed. Scoop out a small amount, and use it, along with a brush, to scrub those hard-to-clean bathroom surfaces. Store any remaining Tub and Tile Scrub in an airtight container.

* Safe for stone

LEMON TOILET PASTE

2 parts borax
1 part lemon juice

Mix borax and lemon juice together to form a paste. Flush toilet to wet sides of the bowl. Use toilet brush to apply paste, allowing it to sit for 1 hour. Scrub with brush and then flush.

Make toilets shine with baking soda, vinegar, and tea tree oil. Just pour, let sit as needed and easily scrub clean.

TEA TREE OIL TOILET CLEANER

1 cup baking soda
1 cup white distilled vinegar
10 drops tea tree essential oil

Combine all ingredients. Pour into the toilet bowl, and allow mixture to sit for at least 15 minutes. Scrub with a toilet brush.

HEAVY-DUTY TOILET SCRUBBER

½ cup borax
1 cup white distilled vinegar
10 drops tea tree essential oil
10 drops of your favorite essential oil

Combine all ingredients and allow to sit in toilet bowl for at least 1 hour. For tougher stains, mix into a paste and apply directly to stains. After it has been allowed to sit, use a toilet brush to scrub toilet clean.

TEA TREE SPRAY

1 teaspoon tea tree essential oil
1 cup distilled water

Mix tea tree oil and water together in a spray bottle. Shake well before each use. Spray in areas where mildew is prone to develop, and allow to air dry. This can also be used to help kill bacteria in toilets or other bathroom areas.

* Safe for stone

PINE FRESH DISINFECTANT

2 cups water
10 drops pine essential oil
1–2 cups white distilled vinegar (optional, avoid with stone)

Combine water and pine essential oil in spray bottle. Spray on the surface and wipe off with a clean cloth. The natural properties of the pine oil will kill any lingering bacteria or viruses. Want more of a kick? Add white distilled vinegar for heavy-duty acidic killing power. Shake well before each use.

* Safe for stone

EASY TOILET FINISH

Want to keep your toilet sparkling fresh and bacteria-free? After you clean the toilet, spray it with hydrogen peroxide, and wipe with a clean, dry cloth.

GROUT CLEANER

To clean dirty grout, mix enough hydrogen peroxide with baking soda to form a paste. Apply to the area. Allow to sit for a few minutes, and then scrub gently with a brush.

* Safe for stone

SCOURING POWDER

2 parts baking soda
1 part salt
1 part borax

Mix together. To use, wet surface with water or vinegar, sprinkle powder on, and let it sit for 5 minutes. Scrub with a brush. Rinse.

* Safe for stone

WHITENING SOFT-SCRUB PASTE

Do you have a stained area that could use a good scrub? Try this naturally whitening paste.

2 tablespoons cream of tartar
Hydrogen peroxide

Mix enough hydrogen peroxide with the cream of tartar to form a paste. Apply the paste to the stained area, let the paste sit for a few minutes, and then scrub with a soft cloth. Rinse well after use. This works well on many stained areas, including those hard-to-get-rid-of tub stains.

* Safe for stone

MOLD AND MILDEW

If you have mildew in your bathroom, start by making sure the room is well ventilated. Give your bathroom's exhaust fan a good cleaning: take off the cover, wash with soap and water, and then use a wet cloth to clean the fan blades. If the blades have mildew, use the Mold and Mildew Cleaner (below) on them.

Open your bathroom window to ventilate the room and allow the sunshine in; this cuts down on dampness and helps prevent mold and mildew. Check for and seal up any leaks around pipes and faucets.

Consider using fabric shower curtains. Vinyl curtains (the ones that contain polyvinyl chloride, or PVC) emit phthalates, those nasty carcinogenic, endocrine-disrupting chemicals linked to a variety of health concerns. Instead of vinyl, look for thin fabric liners or PVC-free liners, either of which will dry quickly to resist mold and mildew. If your fabric curtains show signs of mildew or mold, launder them, and let them dry in the sun. For plastic liners, scrub as best you can with our Mold and Mildew Cleaner and let dry in the sun.

MOLD AND MILDEW CLEANER

2 tablespoons borax
2 cups warm water
¼ cup white distilled vinegar

Add borax to warm water, and shake until borax is dissolved. Add vinegar. Spray on surfaces affected by mold or mildew. Let sit for a few minutes, and then scrub mold or mildew away with a sponge or scrub brush.

> Prevent shower mold and mildew by spraying straight vinegar in and around your shower after each use. Use essential oils for an added antifungal boost or a pleasing aroma.

MOLD AND MILDEW PREVENTION SPRAY

2 cups white distilled vinegar
20 drops tea tree oil or thyme oil (optional)

To help prevent mold and mildew from forming on hard surfaces or plastic liners, keep a spray bottle of straight vinegar in your bathtub or shower; spray liberally after each bath. Allow to air dry or use a squeegee. You can also add 20 drops of tea tree oil or thyme oil to the vinegar; both have antifungal properties. Alternatively, you can add the tea tree oil or thyme oil to water; but vinegar will do more to kill the mold. Using a hydrogen peroxide spray can also help inhibit the growth of mold and mildew.

To help prevent mold and mildew from spreading across the bottom of your shower curtain liners, make certain your shower curtain has plenty of air circulation to dry. After most of the water has dripped off, move your curtain so that it can finish air drying properly. This may mean pulling it to the outside of the tub so it can hang straight or just pulling it across the shower to give the entire curtain and liner a chance to air out.

> **■** Add Thieves Oil Blend (page 58) to hand soap to make your own natural antibacterial soap.

HAND SOAP

With all of this cleaning, you may notice that your family's hands could use some cleaning of their own. Here are some easy solutions.

DISINFECTING HAND SOAP

Thieves Oil Blend hand soap is particularly good to use if your family has been exposed to illness or is fighting a virus.

8 drops Thieves Oil Blend (page 58)
8 ounces of liquid castile soap or Foaming Soap Solution (page 110)

Mix the Thieves Oil Blend with the castile soap and pour into a hand pump soap dispenser. Forget the antibacterial soaps: Fight nature with nature!

MAXIMIZING YOUR SOAP

If you wash your hands often or if you have children, you may notice that you go through a lot of soap. If you are using liquid soap, which doesn't leave that nasty soap scum on your counter, this can get quite expensive. The truth is, if you are using straight soap, you are using way too much.

FOAMING SOAP SOLUTION

3 tablespoons castile soap or other liquid soap
1 cup water
10 drops essential oil (optional)

Mix the liquid soap with the water. Stir well so that all soap is dissolved. Shake if you like, but let the suds settle down before pouring. Fill a foaming soap dispenser with your foaming soap solution and use as normal.

Add as much as 10 drops of your favorite essential oil for an aromatic hand soap that packs an extra antibacterial punch. Alternatively, you can use a scented castile soap with or without adding an essential oil.

FLOORING AND WALLS

Two areas we tend to overlook when cleaning our homes are the flooring and the walls. These areas collect more dirt than we might imagine.

FLOORING

The challenge of keeping a green walkway isn't only the dirt and germs, it's the flooring itself. Carpet and wood floors off-gas nasty volatile organic compounds (VOCs), vinyl floors can contain carcinogenic phthalates, and many different types of flooring (and the material used to tack it down) are treated with or made from chemicals that can be harmful to your health. If you're looking for new flooring, we recommend reading *Understand Green Building*

Materials[125] to educate yourself on the different options and their environmental impacts.

> Leave shoes at the door to reduce the amount of dirt and germs tracked into your house.

Much of the dirt on your floor is tracked in from the outside on your shoes. One of the easiest ways to cut down on dirty, germy floors is to create a no-shoes policy. Make a habit of taking your shoes off each time you come in, and put them on a towel or tray just inside the door.

HARD FLOORS

Although hard floor surfaces are the best flooring for allergy sufferers, dirt can be very visible. Use some of these easy-to-make hard floor cleaners to keep them sparkling clean.

HARD FLOOR SURFACE CLEANER

⅓ cup water
⅓ cup white distilled vinegar
⅓ cup vodka
A few drops dish soap
10–15 drops essential oil

Mix all ingredients, including your choice of essential oil, in a spray bottle. Shake well and spray on floor surface. Wait a few minutes, and wipe up with a cloth wipe or microfiber towel.

CITRUS HARD FLOOR CLEANER

1 gallon hot water
2 tablespoons (or large squirts) castile soap
20 total drops of your favorite citrus essential oils (orange, lemon, lime, grapefruit)

Fill a bucket with hot water. Add soap and essential oils, and mix thoroughly. If using a mop, allow the mop to soak in the water until it is ready to use. Alternatively, use a cloth wipe or scrub brush, depending on your needs, to wash the floor by hand.

FOREST FRESH HARD FLOOR CLEANER

1 gallon hot water
2 tablespoons (or large squirts) castile soap
20 total drops of your favorite wood essential oils (pine, cedar, juniper, cypress)

Fill a bucket with hot water. Add soap and essential oils, and mix thoroughly. If using a mop, allow the mop to soak in the water until it is ready to use. Alternatively, use a cloth wipe or scrub brush, depending on your needs, to wash the floor by hand.

ORANGE WOOD FLOOR CLEANER

1½ cups water
1½ cups white distilled vinegar
20 drops orange essential oil*

Mix water and vinegar together in a small spray bottle. Add essential oil and shake well before each use.

* Another essential oil can be substituted for the orange oil. Consider trying lemon or lemongrass, peppermint and eucalyptus, or ylang ylang for a light floral scent.

Skip toxic mop solutions. Simply use hot water, castile soap, and your favorite essential oils to clean almost any floor in your house.

SUMMERTIME MOP SOLUTION

Clean your hard floors and deter insects with this smell-good mop solution.

1 gallon hot water
1 cup vodka
2 tablespoons castile soap
10–20 drops peppermint essential oil

Mix everything together, wet a mop, and scrub the floor. For an added scrubbing boost for tile, vinyl, or laminate floors, sprinkle baking soda lightly onto the floor before mopping, and scrub with the wet mop.

You can add different essential oils to your mop solution to fit the season or your mood: cinnamon and cloves for winter holidays; pine, tea tree, and lemon for a spring green smell; or sandalwood, lavender, and bergamot for a soothing, relaxing atmosphere.

SCUFF MARK REMOVER

An easy way to remove a scuff mark from a hardwood floor is to place a few drops of cedar oil on the mark. Allow it to sit for a minute or so, then use a clean cloth to gently rub the mark away.

FLOOR WAX

To keep hardwood floors looking beautiful, wax them annually or as needed with this homemade recipe.

⅛ cup beeswax
⅛ cup carnauba wax
1 cup jojoba oil

Melt waxes (use a double boiler or microwave) and add jojoba oil. Mix ingredients together, stirring intermittently as the mix cools. Apply with soft cloth to wood floors; rub and buff until your floor shines.[126]

CARPET

If you have carpet, vacuum frequently with a vacuum cleaner that has a HEPA filter. Without a good HEPA filter, your vacuum is likely kicking the dust and germs right back into the air. Empty your bagless canister outside to minimize the chance of dirt going back into the air inside your house. Be sure to wash your HEPA filter regularly to maximize efficiency.

CARPET FRESHENER

1–2 cups baking soda
10–20 total drops of your favorite essential oils

Lift your mood, combat smells, and deter insects, all while cleaning your carpet. Add 10–20 drops of your favorite uplifting essential oils to the baking soda, giving the combination a good shake. Sprinkle onto carpet and let it sit for as long as 20 minutes before vacuuming as usual.

Here are a few essential oil blends to try:

Refresh your mood and dissolve stress: 10 drops bergamot, 5 drops ylang ylang, 5 drops grapefruit, lemon, or sweet orange

Stinky room fighter: 5 drops lemongrass, 5 drops sage, 5 drops lavender, 5 drops bergamot

Insect deterrent: 10 drops peppermint, 5 drops geranium, 5 drops lavender

CARPET STAIN REMOVER

Spritz vinegar or a 1:1 mixture of vinegar and water on the stain, let sit for a few minutes, and blot with a cloth.

Clean tough carpet stains with equal parts salt, borax, and vinegar. Scrub the stain with the mixture, let dry, and vacuum up. Then blot with vinegar.

HEAVY-DUTY CARPET STAIN REMOVER

1 part salt
1 part borax
1 part vinegar

Mix all ingredients together. Scrub stain with mixture. Allow to dry. Vacuum up any remaining residue. Follow by spritzing with vinegar and blotting with a cloth.

ESSENTIAL FOAMING CARPET CLEANER

2 cups water
½ cup castile soap
10 total drops essential oil of your choice

Place all ingredients in a blender and mix well. Scoop some of the foam off the top of the mixture and place on soiled carpet. Use a clean cloth to rub the foamy soap into the area. Dry thoroughly and then vacuum.

ESSENTIAL CARPET SHAMPOO FOR STEAM CLEANERS

1 cup hot water
¾ cup white distilled vinegar
1½ tablespoons liquid castile soap
20 total drops essential oil of your choice

Combine all ingredients and shake well. Add the Essential Carpet Shampoo to your carpet steam cleaner in place of commercial detergents, and steam clean the carpets as per machine directions, making certain to remove as much liquid as possible. Choose essential oils based on your favorite scents or on your specific deep-cleaning needs.

While wet mopping noncarpeted floors is better than dry mopping (to better capture airborne allergens), avoid "wet-cleaning" or soaking carpet with water, as standing moisture in carpet fibers leads to mildew and encourages bacterial growth. "Dry cleaning" (with the Heavy-Duty Carpet Stain Remover) or cleaning with foam or vinegar are healthier options.

WALLS

What with scuff marks, little hands, and general wear and tear, walls can use a good cleaning every once in a while. Don't forget what's on the walls, especially items that are frequently touched. Light switches and electrical outlets tend to collect dirt on and around the plates. *Carefully* clean around these, so you don't electrocute yourself. Clean hand railings, too. They tend to build up quite a bit of dirt and grime.

> Don't forget to clean your walls, which can collect more dirt than you might realize. Just use an all-purpose solution and your favorite essential oils.

ESSENTIAL VINEGAR WALL SPRAY

1 cup water
1 cup white distilled vinegar
10 total drops essential oil of your choice*

Place all ingredients in a small spray bottle, and shake well before use. Spray any soiled areas and wipe or lightly scrub with a clean, soft cloth.

* Citrus oils work particularly well for cleaning walls and uplifting the spirit.

Wall Marks

Commercial products made for cleaning marks from walls and hard surfaces seem to be magical, and indeed for many families of small children, they are. However, commercial sponges are neither biodegradable nor eco-friendly, despite their lack of toxic cleaning sprays. They also tend to be expensive. Although claims that they contain formaldehyde are false, manufacturers are clear that they should be used with care and that all residue should be rinsed from surfaces. The main ingredient is a formaldehyde-melamine-sodium bisulfite copolymer, which was originally formulated as an insulator and fire retardant, and is not the same as plain formaldehyde. Even though dangers regarding possible chemicals have been exaggerated, manufacturers warn that the sponges and the surfaces where they have been used should not be licked. Additionally, long-term exposure to melamine has been associated with health issues, especially among children.[127, 128]

> There is no need to buy expensive Magic Erasers. Make your own inexpensive wall cleaners using baking soda, water, and lemon oil.

So what is the magic behind this product, anyway? The sponge acts as an abrasive. It essentially buffs and sands away the marks from surfaces. You can do this at home with products you already have in your kitchen.

MAGIC LEMON WALL MARK PASTE

3 tablespoons baking soda
5 drops lemon oil
Water

Mix the baking soda and lemon oil together, adding enough water to make a paste. Apply to mark. Using a clean, soft cloth, gently rub the mark to remove. Please note that because baking soda is an abrasive, it can change the finish on some surfaces.

FURNITURE

Furniture can be one of your biggest investments. It can also see some of the most wear and tear from daily use. You may be tempted to purchase furniture (or carpet or other household goods) marketed as stain-resistant. But the chemicals that make goods stain-resistant are harmful to your health and the environment.[129] Perfluorooctanoic acid (PFOA), the key processing agent in stain-resistant materials, has been linked to both cancer and birth defects, and it is persistent in the environment. PFOA is also used in nonstick cookware, such as Teflon.[130]

Instead of purchasing toxic furniture, clean spills and messes immediately, and do preventive cleaning regularly to preserve the look and feel of your furniture. Keeping furniture clean naturally protects your family's health and prolongs the life of your investment. Use the recipes in this section so that you can rest easy in your easy chair.

CLEANING WOOD

Cleaning and protecting your wood furniture takes only a few minutes and some ingredients from your kitchen. Try these recipes to clean and protect your wood furniture.

JOJOBA WOOD CLEANER

A few drops jojoba oil
½ cup white distilled vinegar
A few drops essential oil of your choice

Mix ingredients together in a spray bottle, and use on wood furniture to clean and disinfect. Scent with your choice of essential oils.

Clean your wood naturally with lemon juice, castile soap, and lemon essential oil.

LEMON WOOD CLEANER

¼ cup lemon juice
1 teaspoon sweet almond oil
5 drops lemon essential oil

This recipe combines the refreshing scent of lemon essential oil with the cleaning power of lemon juice. Put all ingredients in a small spray bottle, shake well, spray, and wipe with a clean cloth. Use immediately or store leftover cleaner in the refrigerator.

OLIVE OIL POLISH

1 cup olive oil
½ cup lemon juice
5–10 total drops of citrus essential oils (optional)

Mix olive oil and lemon juice well. You may need to shake periodically, as the lemon juice and olive oil separate. Spray on wood furniture and polish with a soft cloth. The solution will help moisturize the wood, making it shiny and adding a fresh scent. For a stronger scent, add 5–10 drops of citrus essential oils. Use immediately or store leftover cleaner in the refrigerator.

COCONUT-ALMOND WOOD POLISH

⅛ cup melted coconut oil
¼ cup sweet almond oil
1 teaspoon almond extract

Combine melted coconut oil with almond oil. Add almond extract and mix well. Using a clean, soft cloth, rub the polish into wood. Wipe off any excess with a dry cloth wipe.

> Easily remove water rings from wood furniture with olive oil or peppermint essential oil.

PEPPERMINT WATER RING REMOVER

To remove the water rings caused by condensation on glasses from wood furniture, place a few drops of peppermint essential oil on the stain. Allow to sit for a while and then buff with a clean, soft cloth.

OLIVE OIL WATER RING REMOVER

To remove water rings using olive oil, cover the stain with an adequate amount of olive oil. Let the oil sit for several hours. Use a soft, clean cloth to buff the area and remove excess oil.

FABRIC FURNITURE AND ACCESSORIES

Keep your upholstery and curtains dust-free by regularly vacuuming them with your brush attachment. When you do need to clean up a spill or stain on a fabric-covered cushion, do not douse with water, because the water can remain in the padding after it dries on the cover. Spray lightly and allow to dry (in a sunny area, if possible).

UPHOLSTERY AND CURTAIN SPRAY

¼ cup vodka
¾ cup water
½ tablespoon castile soap

Combine all ingredients in a spray bottle. Spray the soiled area of the upholstery or curtains, and gently work the solution into the spot with a clean, soft cloth. Reapply as necessary until the area is clean. Shake well before each use.

LEATHER PROTECTION

To keep leather from cracking, it needs to be moisturized and protected. Our Coconut Leather Protector does both.

Protect leather furniture with beeswax, coconut oil, and olive oil.

COCONUT LEATHER PROTECTOR

1 ounce beeswax
½ cup coconut oil
½ cup olive oil

Carefully heat the beeswax and oils, and mix well (use a double boiler or microwave). Remove from heat and pour into a glass jar. Allow to cool and then store with an airtight lid. To use, apply the leather protector with a soft cloth, wiping off any excess.

MATTRESSES

You won't want to forget to clean your mattresses. You may have heard the urban myth that mattresses will double in weight in a ten-year span. Although you won't break your back trying to flip your older mattress, it is true that mattresses do get dirty.

As you sleep, skin cells slough off onto your mattress. Those cells attract microscopic insects called dust mites. You can't see these tiny creatures, but they are there, along with their cousins, children, dead relatives, and fecal matter.[131] For those with allergies or asthma, dust mites can be a major problem, especially because we spend an estimated one-third of our lives sleeping among them.

Use dust mite covers on mattresses to help minimize allergies.

Dust mites on mattresses can be minimized by using dust mite–proof covers. These covers completely encase the mattress. Many of these covers are made of plastic or vinyl, but you can find ones that are made of fabric. Just make certain that the label says they are for allergies or for dust mites. The tiny creatures can go through spaces in between fibers, so you want to be certain your cover has spaces they can't make it through.

Regardless of whether you have an allergen cover, you will want to clean your bed and bedding on a regular basis.

TIPS TO CLEAN YOUR BED

Remove all of the bedding. This is the first step and one that should probably happen on a weekly basis. Give it all a good wash in hot water. Hot water will kill dust mites, whereas cold water will not. You may choose to go longer between washing pillows, but don't forget that they collect dirt and matter, too. Most pillows, including synthetic, cotton, and down-filled pillows, can be washed. To prevent pillows from turning into hard lumps, throw a couple of Wool Dryer Balls (page 146) into both the washer and dryer along with your pillows. If you have a wool pillow, avoid the chemicals from dry cleaning. Freeze the pillow for 24 hours to kill any dust mites, spot clean, and allow it to sit in the sun to freshen the pillow.

Vacuum your mattress. Make certain you lift your mattress and vacuum in between the mattress and the box springs.

Using baking soda with some lavender oil in it, sprinkle the surface of the mattress thoroughly, and let it sit for an hour. Old shaker containers work well for this.

Pull the vacuum back out, and vacuum up all of the baking soda. The baking soda helps draw out any moisture, along with dirt and grime.

WASHING MACHINES AND DRYERS

Obviously, the outside of your washer gets dirty. Ditto for your dryer. A few spritzes of an all-purpose spray (page 56) and a clean cloth can take care of that, but what about the inside? It needs maintenance, too.

CLEAN YOUR WASHING MACHINE

To clean the inside of your washing machine, run a cycle on hot water with a quart of vinegar and a cup of baking soda. If your machine is smelling particularly musty — which is a strong possibility, because most of us tend to keep the washing machine door shut when not in use — try adding 10–20 drops of tea tree oil.

> Prevent your washing machine from becoming musty by leaving the door open and allowing the machine to air dry after use.

Don't forget to check the inside of the door and any gaskets. They can easily have residue buildup and should be cleaned. Use some all-purpose spray and a clean cloth to clean any grime or buildup.

DRYER LINT

> Lint buildup in dryers causes more than 15,000 fires every year.

You know to clean out the lint trap in your dryer, but do you realize the lint trap doesn't catch all of the lint? On a regular basis — at least once a year — take apart the dryer vent and give it a thorough cleaning. Not only will this make your dryer more efficient, but you will decrease your risk of a lint fire.

ELECTRONICS

Avoid expensive and toxic electronic cleaners. Most electronics can be cleaned using distilled water on a soft cloth (do not spray the cleaner on the electronics themselves).

From TVs to touch pads to cell phones, all electronics need to be cleaned of dust and fingerprints regularly in order to operate efficiently. Read the packaging for your electronics to see what cleaners are recommended by the manufacturer. Normally, distilled water and a soft cotton cloth will be enough to clean most electronic equipment (distilled water does not have the minerals in it that tap water does and will not leave a residue behind). Microfiber cloths work well for a streak-free shine.

If you need something more, try a 1:1 ratio of vinegar and distilled water or a 1:1 ratio of rubbing alcohol and distilled water in a spray bottle. Spray onto the cloth (not onto the screen or any exposed electronic parts), and gently wipe clean.

For keyboards, air vents, and other electronic nooks and crannies, use canned air to blow away dust and debris.

VAPOR STEAM CLEANERS

Vapor steam cleaners work by forcing steam onto surfaces, including between fibers, which are generally difficult to get between. The steam loosens dirt and debris, making it easier to remove. The superheated steam uses very little water and electricity while sanitizing, cleaning, and deodorizing surfaces.

> Invest in a steam cleaner to clean countless areas of your home with hot water alone — no chemicals needed.

BENEFITS OF STEAM CLEANING

- Steam forgoes the use of toxic and expensive chemicals, instead using plain water via steam to clean.

- Steam cleaners leave behind no residues and don't stain clothing or fabric.

- The high heat used in steam kills bacteria such as E. coli, listeria, and salmonella.

- Because they use only water, steam cleaners are environmentally friendly.

- Steam cleaners are highly versatile, cleaning a wide range of surfaces in your home.

- Steam cleaners kill what you don't see: bacteria, viruses, and dust mites.

PURCHASING A STEAM CLEANER

You can find steam cleaners in a wide range of prices, from $50 to hundreds of dollars. What should you consider when making your purchase?

BOILER CAPACITY: The boiler is the cavity that holds the water and steam. The larger the boiler capacity, the more water it can hold. So should you go for the largest boiler capacity you can afford?

Not necessarily. A larger boiler capacity also means that it will take the unit longer to heat up.

TOOLS AND ATTACHMENTS: More tools and attachments for an item mean more versatility in cleaning, right? Yes, but if you won't use them, they won't do you any good. Look for tools and attachments you will actually use. Small scrub brush attachments can make it easier to get baked-on or built-up grime off your surfaces. Squeegees may be beneficial if you plan to use your steamer to clean windows, although you may already have a squeegee at home. Evaluate the tools and attachments in the set to see if they will actually add value for you.

STORAGE: How easily does the cleaner store? Is there a way to wrap up the cord? After a few uses, you may get tired of having a cord that just hangs and gets in the way when not in use. Are there easy places to store attachments?

MOBILITY: When working with a steam cleaner, you may move around a lot. Check to see how easily the steam cleaner moves with you. Is it on wheels? Does it pull easily? How portable is the unit? Your steam cleaner should be doing the work, not you.

EASE OF USE: How easy is the system to use? Hand-held units may prevent you from being able to clean intricate areas well. Check to see how the steam is released. Look for units that use a simple pressure trigger rather than a pump-and-release system. Is it easy to refill the boiler?

PRESSURIZED VERSUS NONPRESSURIZED: Another consideration when purchasing a steam cleaner is whether to get a pressurized unit or a nonpressurized unit. The term *pressurized* is misleading. Pressurized units can reach as much as 70 pounds per square inch (PSI), depending on the unit. When compared with a power washer, which typically reaches 1,400 PSI, you realize that pressurized units aren't all that powerful, considering the price.

There is a benefit to pressurized steamers: They give an initial burst of hotter, more powerful steam. This is short-lived, generally lasting only a few seconds, and then you must wait until you stop using the steam and the pressure once again builds up to use that feature. Does a nonpressurized unit have enough pressure to work? Yes, the process of water turning into steam is enough to force the steam out of the unit.

STEAM CLEANERS: NOT CHILD'S PLAY

Steam cleaners are amazingly easy to use and kill bacteria and viruses without the use of chemicals. But don't hand one to your child. Water boils at 212 degrees Fahrenheit, at which point it turns to steam. The steam coming from your cleaner, as well as any metal attachments or nozzles, is very hot. Work safely and enjoy the ease of cleaning with your system.

LAUNDRY SIMPLIFIED

There are four aspects of cleaning your clothing, freeing it from dirt, grime, and any viruses or bacteria. First is mechanical action. Whether you are scrubbing an item by hand or using a washing machine to agitate the clothes, mechanical action helps to remove dirt from fabric. Second is time. It would be nice to be able to clean clothes in no time at all, but anyone who has woken up to find the outfit he or she intended to wear sitting in the dirty clothes knows it is an unrealistic dream. It takes time to wash clothing. Third is chemical reactions. From pretreaters to laundry soap or detergent, the chemicals you use, toxic or natural, help clean your clothing. Fourth and last is temperature. Higher temperatures help to sanitize and clean clothing.

Mechanical action, time, chemical reactions, and temperature may all act singly on the clothes you are cleaning, but together they join forces in the cleaning process. If you decrease the potential cleaning of one force, you need an increase in one of the others.

SORTING LAUNDRY

The first step in washing clothing is sorting. It can be tempting to toss everything in the wash together, but different items have different needs. By sorting your laundry, you can use the correct amount of soap and the correct water temperature to wash your clothing while lengthening the life of your clothes.

> To prolong the life of your clothes, sort laundry according to color or use.

To keep your clothing in good shape, sort it according to color and use. Separating colors allows you to prevent colors from running, and separating types of clothing gives you the opportunity to treat each wash load appropriately. You can use tea tree oil with musty-smelling towels without treating all of your laundry. Heavy fabrics won't be agitated against your delicates, forcing you to spend more money on frequent replacements.

WASHING

Getting clothes clean means removing stains and odors associated with dirt. For the best results, you should treat heavy stains and odors prior to washing. Then use the appropriate washing method and your homemade laundry detergent. Don't be fooled by commercial products with heavy, toxic fragrances. Clean doesn't have a smell. Unless you are adding a product, such as all-natural essential oils, to your laundry, your freshly washed clothes shouldn't smell like anything.

TREATING STAINS AND ODORS

Your best bet for treating a stain, regardless of the type of stain or what it is on, is to take care of it right away. Fresh stains are easier to remove, and you won't find yourself working so hard and getting so frustrated.

Remove any excess liquids or solids by blotting them with a clean towel or gently scraping them off. You want the stain minimized before working on it so that you aren't merely spreading the stain. Treat the stain, but be certain you don't set the stain into the fabric. Some items, such as bar soap, may actually make your problem worse.

Check clothing before washing. We often don't realize that clothing has stains and end up missing the best opportunity to treat them. Give clothes a glance just before you place them in the washing machine.

Check clothing before drying. If you are using a dryer, this is your last shot at treating the stain. Once the clothing goes into the dryer, you lose the ability to treat the stain, as the heat will likely set it.

Isolate the offending piece. If the stain is one that could come off and stain other clothing, you are better off washing the piece separately.

Hydrogen peroxide

Hydrogen peroxide (H_2O_2) makes a great spot remover for organic stains. It is especially helpful in removing blood stains from clothing. Add a small amount of H_2O_2 to the stain and allow to sit for a minute. As the H_2O_2 reacts with the stain, you will notice bubbling. This means it is working. For larger stains, you may need to add more or gently rub the clothing. Wash as usual. As hydrogen peroxide can alter some fabric dyes, test in an inconspicuous area first.

Bleaching Stains

Some stains on white or light-colored clothing will benefit from bleaching. You don't need to get out the chlorine; just turn to nature for a little help. Lemon juice is a natural bleaching agent. You can use it as a prelaundry treatment or add some to your wash. As it does bleach, you may want to test your clothing before using this method. Another all-natural method is to hang the item out in the sun. Sunlight works wonderfully at removing stains and killing bacteria.

Mineral Remover

Minerals are known to be laundry boosters, brightening clothing and preventing the characteristic graying seen in some white fabrics. Adding ½ cup of borax or washing soda to a load of clothing and washing in warm to hot water will help keep your whites white and your brights bright. This method should not be used on clothing that may shrink in hotter water temperatures.

Washing Soda

Washing soda works well to remove many stains. Add ½ cup to your wash, or make a paste to use on difficult stains.

Glycerin

Many older stains are set into the fabric and difficult to remove. Try soaking the stain with glycerin to soften it prior to washing.

Vodka

Applying vodka to an ink stain will help loosen the stain and save the garment from ruin.

Tea Tree Oil

With repeated exposure to moisture, towels can take on a musty smell. If you find that you need to wash towels frequently to avoid that musty smell, try adding 10–20 drops of tea tree oil to your washing machine load and wash as you normally would. The natural antibacterial and antifungal properties of tea tree oil will have your towels smelling fresh again.

Add tea tree oil to your wash to prevent the musty smell towels sometimes develop.

COLOR CATCHER

1 tablespoon washing soda
1 cup hot water
White fabric

Making your own color catchers to remove dyes from the wash is easy. Mix the washing soda into the hot water. Soak pieces of white fabric in the solution and allow to dry. The next time you need a color catcher in your laundry, toss one in.

WATER TEMPERATURE

Choosing the correct water temperature for your laundry is more important than you might think. It may be tempting to cram a lot of laundry together in one load and wash it at the same temperature, but just as with sorting, the water temperature can play a significant role in both the cleanliness of your laundry and its longevity.

HOT WATER

Hot water is used for items that need more cleaning, as the heat helps clean heavily soiled items while sanitizing clothing containing bacteria or fungus and killing insects.[132] This is especially important for white clothing, as it is more likely to show dirt, and with particularly sweaty, oily, or soiled items. Items such as towels, which can harbor mold and mildew, can benefit from hot water washing, as can cloth diapers, as they are heavily soiled and may contain bacteria. Hot water isn't good for all laundry, as it can fade colors, shrink natural fabrics, and set some protein stains into fabric.

COLD WATER

Cold water is the most energy-efficient water temperature for laundry. It is less likely to shrink clothing than hot water and works well for many dark or brightly colored clothes, as it doesn't promote fading.

Cold water won't sanitize your clothes, though, and it is less efficient at removing stains. Pretreating stains helps remove stains on clothing washed in cold water. Choose liquid detergent or soap when using cold water, as it is more readily dissolved than its powdered counterpart.

WARM WATER

Warm water is the middle ground of washing. Like hot water, it helps to dissolve powdered detergents — and with greater energy efficiency than hot water. Due to higher heat, warm water is more likely than cold water to fade clothing. It does not sanitize clothing and does not work well at removing heavy stains and soils. It tends to be best used for lightly soiled clothing and synthetic fabrics, such as nylon, polyester, spandex, and rayon blends.

READ THE LABEL

The garment label will tell you what temperature is recommended for the garment. If you are in doubt regarding proper care instructions, use cold water. You can always try again with warm or hot water if the item doesn't get clean enough.

THE RINSE CYCLE

Generally speaking, regardless of what temperature you use to wash clothes, most clothes should be rinsed in cold water. Rinse water has little effect on either sanitization or stain removal. Using cold water is energy-friendly and budget-friendly.

DETERGENTS AND SOAPS

Commercial laundry detergents can be expensive, and even the fragrance-free versions tend to have some sort of perfume. You can avoid fragrances and other toxic chemicals by making your own laundry soaps from easy recipes.

BASIC LIQUID LAUNDRY SOAP

2¼ cups liquid castile soap
1 tablespoon glycerin
¾ cup water
¼ cup white distilled vinegar
10–20 drops essential oils of your choice

Combine all ingredients. Shake well before each use. Add ¼ –½ cup of solution, depending on how heavily soiled the laundry is, to each load of wash.

BASIC BULK LIQUID LAUNDRY SOAP

1 quart boiling water plus 2 gallons room temperature or warm water
2 cups soap flakes (or grate your favorite all-natural soap)
2 cups borax
2 cups washing soda
10–20 total drops essential oils of your choice

Bring 1 quart of water to boil. Add soap flakes or grated soap, stirring until melted and the soap is incorporated into the water. Remove from heat. Carefully add borax and washing soda and stir well. Add the remaining water, stirring until a uniform consistency has been reached. Allow to cool before adding essential oils. Store in a covered container, stirring gently before use. Use ¼–½ cup of soap per load of laundry.

BASIC POWDERED LAUNDRY SOAP

1 cup washing soda
1 cup baking soda
2 cups soap flakes (or grate your favorite all-natural soap)
5–10 drops essential oil (optional; skip this if you have chosen a scented soap)

Combine all ingredients thoroughly, and store in an airtight container. Use ⅛–¼ cup per load of laundry.

SOAP NUTS

As an alternative to laundry detergent or soap, many people are turning to soap nuts, also known as soapberries. Soap nuts come from the genus *Sapindus*, which consists of multiple small trees and shrubs that are native to temperate and tropical regions. The drupes, or fruits, of the species contain saponin, a natural surfactant.[133] Due to their natural sudsing capabilities,[134] soap nuts are used in various all-natural soaps.[135]

Soap nuts have been used for thousands of years by indigenous peoples in the areas where the plants naturally grow.[136]

Tips for Purchasing Soap Nuts

Purchase by weight. How many loads of laundry you will be able to wash with soap nuts varies, based on washer type, washer size, cycle duration, load size, water hardness, water temperature, size of the soap nuts, saponin concentration, and more. Avoid purchasing soap nuts from companies that advertise the number of loads they will wash as a way to sell their berries. Although reputable companies may give rough guidelines to give you an idea of what to expect, they will sell their soap nuts based on weight.

Take a look at the company's claims regarding loads. On average, ½ ounce of soap nuts (approximately five average-sized, deseeded soap nuts) will clean anywhere from three to seven loads, depending on the factors already mentioned. If the company claims an outrageous number of loads washed, look for a different brand.

The only way to accurately compare the price of soap nuts is to compare price per ounce.

Buy only deseeded soap nuts. The seeds do not aid in washing. Besides doubling the weight of the soap nuts and thus costing you more, seeds left on wet laundry can permanently stain clothing. Look for soap nuts that specifically say "deseeded." Watch out for companies that are not explicit about this or that claim they are selling "whole soap nuts."

All soap nuts are organic, regardless of whether they are advertised as such. A company advertising organic soap nuts is probably doing so to justify a higher price. There is nothing special about their soap nuts, as all are grown without the use of chemicals. Is there an exception to this? Yes. Official organic certifications include USDA Certified Organic and EcoCert Certified Organic. These certifications involve strict regulations and assure consumers of chemical-free and sanitary processing.

Buy *mukorossi* or *trifloliatus* soap nuts, as they are high in saponin. Mukorossi have a higher market value, so if you are buying trifloliatus, you should be paying a significantly lower price. The problem with trifloliatus is that it is very similar in appearance to its lower-saponin-containing cousins. Mukorossi, on the other hand, has a distinctly different look. Read up about what you want before buying. Soap nuts are a buyer-beware industry.

FABRIC SOFTENER

Commercial fabric softeners are not only toxic, they damage your clothing. Fabric softeners coat the surface of fabric with a thin layer of lubrication. This makes your clothes feel smoother and helps to prevent static buildup. This same process reduces the ability of fabric to absorb liquid, making items such as towels and cloth diapers less efficient. It also means that every time you put on your clothes, you are putting on a layer of toxic chemicals and fragrances. You can easily replace your commercial fabric softeners with natural products to make your clothes softer and less likely to have static cling.

Soften clothes naturally by adding baking soda or vinegar to each load of laundry.

EASY BAKING SODA FABRIC SOFTENER

Baking soda will naturally soften laundry. Add ¼–½ cup of baking soda to your wash. Amounts will vary due to local water hardness. If you have harder water, you may need to increase the amount of baking soda.

Vinegar

Vinegar also makes an excellent fabric softener. It is nontoxic and strips away the residue left on clothing by soaps and detergents. Although the vinegar smell will wash out in the rinse cycle, you may want to add a slight scent to your laundry. Scenting the vinegar with essential oils can give a pleasant, light aroma to your clothing without the harsh chemicals and synthetic perfumes used in many commercial products.

ESSENTIAL FABRIC SOFTENER

1 gallon white distilled vinegar
20 drops essential oil of your choice*

Add essential oil to white distilled vinegar and store in a sealed container. Add up to 1 cup of solution to your wash cycle.

* Possible choices include lavender, sweet orange, peppermint, lemongrass, ylang ylang, eucalyptus, lemon, and rose geranium.

DRYER SHEETS

Have you stopped using commercial dryer sheets but find that you miss having your clothes scented? Try these all-natural Quick and Easy Dryer Sheets. Your clothes will come out lightly scented.

Use essential oils to naturally scent your laundry.

QUICK-AND-EASY DRYER SHEETS

4 or 5 drops essential oil of your choice
Scrap fabric (cotton works well), washcloths, or unmatched socks

Place drops of essential oil on a scrap piece of cotton fabric and toss in the dryer.

WOOL DRYER BALLS

Felted Wool Dryer Balls are an easy-to-make, nontoxic alternative to commercial fabric softeners and dryer sheets. Toss three to five of them into the dryer with your load of laundry, and they will soften clothes without using chemicals. If you scent the balls with a few drops of essential oils, they will lightly scent your clothing. Wool Dryer Balls also help to lessen static and, according to some reports, lessen drying time, so that you use less electricity or natural gas to dry your laundry. And because these handy orbs can be used indefinitely, you save money.

Replace toxic dryer sheets with economical and reusable Wool Dryer Balls.

The balls can be purchased from numerous companies, from large commercial companies to work-at-home moms. However, you can also easily make them yourself. A set can be a lovely gift for someone who is interested in becoming more environmentally friendly, or for someone who is interested in saving money, but needs a little push.

If you are going to make your own wool dryer balls, there are a few things to keep in mind. The key is 100 percent wool, as you want the yarn to felt. In the felting process, the fibers bleed into each other, forming a thick fabric just like a sheet of felt. Look at the wool's label. It should say something about requiring hand washing. That is exactly what you want for this project, although you won't be hand washing your Wool Dryer Balls. You are going to be making a solid yarn ball, so feel free to use cheaper wool yarn on the inside. If you want a prettier, more expensive yarn, save your money and use that only for the outer layers, where you can see it. You may find some wool that says it is wool roving or felting wool. This is generally more expensive but can make some dramatic-looking Wool Dryer Balls. Again, save your money and use this for the outer layers.

WOOL DRYER BALLS

100 percent wool yarn
A sock (an old sock is fine)
Cotton string
Essential oils (optional)

Start by making a ball out of your wool. If you are using two different kinds of wool, be sure to use the inner color for this. Wind your ball tightly. You don't want it unraveling before or during the felting process. Continue winding your ball of yarn evenly until you have a ball that is approximately the size of a tennis ball. Securely tuck in your yarn end so that it doesn't unroll. This will be the core of your Wool Dryer Ball.

Next, place your newly wound ball into the sock and tie securely with the cotton string. A tighter sock will hold the shape of the ball better. The sock is not permanent. After the felting process, it will be taken off. You just want your dryer ball to keep its shape until it is felted.

Felt the core. The most efficient method is to use a washer and dryer. Sure, you could felt your wool by hand at the sink, but if time is of the essence, go for the easy method. This is your chance to abuse the wool yarn, which is usually treated with gentle care. Wash it on hot in the washing machine. Use soap and make certain you have some agitation. Try washing it with heavy items that also need hot water to be sure it gets some rough treatment. Then throw it in the dryer and put it on high. Feel free to throw it in for multiple cycles with your clothes to be certain it felts properly. This is an important step, and you don't want to skimp.

Take your newly felted wool ball and add more wool to it. If you are using a different yarn or roving for the outside, this is the time to use it. Wrap your inner ball completely with the new yarn until it is slightly larger than a tennis ball or to your desired size. Once again, securely tuck away the end of the yarn.

Place the ball back in the sock and tie it. Repeat the felting process by washing with hot water and drying on high heat. You want the outsides of your Wool Dryer Ball to be thoroughly felted, so feel free to repeat. If you plan to add essential oils to your dryer ball, do so at this time. Choose from your favorite scents and add just a few drops to each ball. You don't want to saturate the ball.

CLOTHESLINES

For generations, clotheslines have been used to dry clothing. Clotheslines are a frugal choice, with no costs after the initial small investment.

> Dryer lint doesn't magically appear out of thin air. The lint you clean from your dryer's lint trap is actually fibers from your clothing.

They are environmentally friendly, using no fossil fuels and emitting no greenhouse gases. Natural drying is better for clothes. The high heat of dryers is hard on your clothing and can cause some fabrics to shrink. During mechanical drying, small fibers are removed, damaging your clothing over time and potentially reducing air quality from the airborne lint. Static is not an issue for clothes that are air dried. You get that inimitable clothesline-fresh smell.

Not everyone has the space for a clothesline. An alternative is an indoor drying rack. Many styles can be purchased or made, ranging from high-tech racks that pull down from the wall or ceiling to inexpensive stand-alone racks that fold up when not in use.

STARCH

If you have an item that needs to be starched, you can quickly make your own all-natural starch without the need for aerosol cans or the chemicals that come with them.

HOMEMADE STARCH

2 teaspoons plant-based starch (such as cornstarch, potato starch, or arrowroot starch)
1 cup water

Completely dissolve starch in water. Pour into a spray bottle. Lightly spray clothing before ironing. Use sparingly to avoid leaving a starch residue on the clothing.

STORING OUT-OF-SEASON CLOTHING

A lot of things can happen to clothing while it is in storage. Mysterious stains can appear. Insects and rodents can attack. There are a few things you can do to help protect your stored items, without resorting to commercial chemicals.

Store Clean Clothing

Store only clean clothing to avoid causing stains or attracting pests.

It makes sense not to store dirty clothes, but sometimes dirty items still make it into the bin. Perhaps you wore an item only for a couple of hours and then hung it back up. If in doubt, wash the item before storage. Dirt on clothing can cause some of those mysterious stains to appear through oxidation, and it can attract pests.

AVOID STARCH AND FABRIC SOFTENER: Starch makes an attractive snack for insects. Wash any items that have been starched to remove the tempting treat. Similarly, avoid using fabric softener before storing clothing.

REMOVE PACKAGING: Remove any packaging touching your clothing. Hangers with foam or velour coverings can stain your clothing. Remove the thin plastic bags, such as those from a dry cleaner, to help avoid mold from residual moisture.

FREEZE IT: If you think that certain items might have been subjected to moth larvae — wool sweaters, for instance — try freezing the items to kill any bugs. This way, you won't be storing larvae with a convenient food source.

MAKE IT AIRTIGHT: Bugs and rodents, along with many bacteria on your skin, need air to breathe. Placing items in airtight containers helps prevent accessibility. Airtight cedar chests are perfect for storing clothes, as they also have the natural cedar oil to repel pests. If you don't have cedar chests, consider plastic storage containers with a tight seal. Cardboard boxes, although generally free, invite pests to chew through and attack your clothing.

CHALK IT UP: Concerned about moisture ruining your clothing? Try adding some chalk to your storage. It will absorb any moisture to help keep mold and mildew at bay.

ADD INSECT DETERRENTS: Although taking precautions in how you store your clothing is important, everyone can use a little more assurance. Try placing some natural deterrents in your storage containers for maximum benefit.

> Use essential oils instead of toxic mothballs to naturally keep insects away from stored clothing.

- *Cedar Balls, Blocks, or Essential Oil:* Cedar oil naturally repels pests. Place multiple balls or blocks of cedar wood throughout your storage container. If they have been used for many seasons and the scent has faded, use sandpaper to sand off the outer layer. Alternatively, use cotton with 4 or 5 drops of cedar oil.

- *Lavender Sachets:* Lavender sachets, made with dried lavender buds or with the more concentrated lavender oil, are perfect additions to your stored clothing. Lavender oil is a natural insect repellent. It is also antibacterial and antifungal.

CHEMICALS IN YOUR CLOTHES

You might not realize that chemicals are used in the production of clothing. Following are some examples.

ANTIBACTERIAL PRODUCTS: Some manufacturers of clothing, especially exercise clothing, apply antibacterial products to their items to inhibit bacterial growth and odors. One example is triclosan, a skin irritant[137] and a known endocrine disruptor.[138] It enters the bloodstream via contact with the skin.[139]

FUNGICIDES: Fungicides are used in clothing primarily to prevent mold during transport. Dimethyl fumerate is a fungicide commonly used in clothing. It can cause strong skin reactions and is banned in the European Union in consumer products.

AZO DYES: Valued for their low cost and vivid, clear colors, azo dyes are a large group of substances that contain azofuntions. Many of these chemicals decompose into mutagenic and carcinogenic arylamines.[140] Due to the water solubility of most of these dyes, they are easily absorbed via skin contact or respiration.

DISPERSE DYES: Disperse dyes are used to dye synthetic fibers. Many of these dyes are allergenic.

FORMALDEHYDE: Formaldehyde is used on clothing to reduce shrinkage and wrinkles, as a dye fixative, or to remove dirt in dry cleaning. It causes skin irritation and allergies[141] and is classified as a known human carcinogen by the U.S. National Toxicology Program.

PHTHALATES: Phthalates are plasticizers and can be taken into the body through respiration or via skin absorption.[142] Children are at particular risk with phthalate exposure. Higher incidences of asthma and allergies are associated with higher levels of phthalates in the home. Many phthalates are endocrine disruptors and may affect fertility.

NONYLPHENOL ETHOXYLATES: Nonylphenol ethoxylates (NPEs) are used in the washing and dyeing of textiles. NPEs break down into nonylphenol (NP), a persistent and bioaccumulative product that is extremely toxic to aquatic life and is thought to impair fertility.[143] NP is not treatable by effluent treatment plants and finds its way into water systems.[144]

HEAVY METALS: Heavy metals are used to dye clothing. Compounds containing heavy metals are toxic and dangerous in high amounts and can cause allergies. Many are carcinogenic.

PERFLUORINATED COMPOUNDS: Perfluorinated compounds (PFCs) are used to make garments stain-, oil-, or water-resistant. PFCs, which encompass a complex group of chemicals, are known to cause reproductive and endocrine problems.

NATURAL FABRICS

Unlike their synthetically made cousins, many of which are made from petroleum products, natural fabrics are made of fibers originating from plants or animals. Because the raw materials are grown, they are easily renewable. However, natural fabrics are not necessarily free of chemicals, and their production is not necessarily ecologically sound. When choosing clothing or fabric, take a look at how the fabric was grown.

CONVENTIONAL COTTON: Toxic chemicals are often used on fibers as they are grown. Although cotton uses only 2.4 percent of the world's agricultural acreage, it accounts for 25 percent of the world's pesticide use! Herbicides are used during harvesting to defoliate the plants and make picking easier. More chemicals are used to process the fibers: bleaching, sizing, dyeing, and so on, require the application of toxic chemicals, many of which are applied with heat to bond them to the cotton fibers. Despite multiple washings throughout the production process, residue remains on the final product.

ORGANIC COTTON: Organic cotton is grown and produced without chemicals.

GMO seeds and chemical pesticides are not used in the production of organic cotton. To control pests, trap crops planted nearby are used to lure away predatory insects, and beneficial insects are used to prey on those that would eat the cotton. Weeds are removed by hand. Synthetic fertilizers are avoided; instead, crop rotation is used to increase organic material in the soil. Picking is usually done by hand after a seasonal freeze naturally defoliates the plants.

Organic fibers are processed differently, with little or no chemical finish. Colored organic fabrics might be dyed with natural dyes or grown in natural colors.

It takes one-third of a pound of pesticides and fertilizers to produce enough conventional cotton to make one T-shirt. That same shirt requires 713 gallons of water to make. Traditional wet-dye methods use an additional 75 gallons of water per pound of fabric.

WOOL: Wool is made from the fleece of animals such as sheep, making it a renewable resource. However, this renewable resource comes with a price. Sheep attract pests, such as lice and ticks, and commercial sheep farms dip their sheep in insecticide baths. Organophosphate insecticides can harm the central nervous system of workers dealing with the sheep and the cut wool. The insecticides leave a residue on the final wool product, which is turned into clothing.[145]

Wool scouring, which removes dirt and lanolin,[146] requires a large amount of water and, for commercial operations, a large amount of chemicals. This results in heavily polluted wastewater. Further treatments, such as dyeing or mothproofing, involve highly toxic chemicals, which create toxic wastewater and leave more residue on clothing.

WASHABLE WOOL: Many people shy away from wool, a fabric that must be handled with care. Who wants to purchase a garment that shrinks after being washed or dried improperly? But with the advent of washable wool, consumers are once again turning to wool. It is used for everything from soft sweaters to diaper covers.

Is it safe? Almost all washable wool goes through the Chlorine-Hercosett process, which removes the scales of the natural wool fiber, modifying it into a smooth, synthetic-like fiber. A strong acidic chlorine solution is used, and then a polymer resin is applied. The effects of long-term exposure to machine-washable wool are not known, but we do know that the chlorine waste products are highly toxic to aquatic life in low concentrations.

Is it wool? Sure, the final product is still wool, but the wool fibers are now encased in a plastic resin. Those lovely qualities you associate with wool — warmth and water protection — are gone.

SUSTAINABLY RAISED, ORGANIC WOOL: Sustainably raised wool comes from sheep nourished with a diverse feed crop, rather than a single feed crop. This is a more easily digestible feed for woolly animals, lowering the emissions of greenhouse gases. Farmers do not use pesticide dips, and they use low-impact processing for the sheared wool.

SILK: Silkworms have been cultivated for their silk for more than five thousand years.[147] Silk is a natural fiber, created with the silky strands that silk moth caterpillars use to make their cocoons. In traditionally raised silk, hatched caterpillars eat mostly mulberry leaves and increase in size until they are ready to make a cocoon. Fewer pesticides are used on the mulberry trees than are used for other fabric crops, as the silk moth caterpillars are sensitive to them. When the cocoon is finished, it is dipped in hot water or steamed, killing the caterpillar inside and beginning the unraveling of the silk filament.

> Silk cocoons are made from one continuous silk thread, which can be anywhere from 1,000 to 3,000 feet long.

PEACE SILK: Some companies now make peace silk. The producers of peace silk wait for the moths to emerge from their cocoons before salvaging the now-damaged cocoons for silk production.

WEIGHTED SILK: Weighted silk is silk that has had metallic salts added to the fabric to increase the body, luster, and weight of the fabric. The metals used to weight the silk — often chromium, barium, lead, tin, iron, or sodium magnesium — not only damage the strength of the final product but can cause health problems for the wearer.

FINISHING SILK: Silk easily creases and wrinkles, so many clothing manufacturers apply softeners, elastomers, and synthetic resins to the fabric to increase its antiwrinkling properties, causing adverse effects for workers and the environment.[148] Other chemical treatments may be used to repel water and oil, increase flame retardation, or make the garment easier to care for. Textile acid dyes used to dye silk requires high levels of chemicals, resulting in large amounts of contaminated wastewater. Many of the chemicals used in this process are classified by the EPA as having a moderate to high level in regard to concerns regarding carcinogenicity.

ORGANIC, ETHICAL, OR SUSTAINABLE SILK: When buying silk, what is your main concern? If you want to limit chemical exposure, look for raw, undyed silk or raw silk dyed with a low-impact, fiber-reactive dye. There is currently no formal standard for what constitutes organic silk, so be wary of the label. If you are concerned about the ethics of silk, look for wild silk or spun silk. Regardless, look for silk from fair trade companies, which protect the workers involved in all phases of silk production.

HEMP FIBER: Hemp is one of the earliest known domesticated plants and has been cultivated for more than twelve thousand years.[149] The fiber is very strong and durable, with a texture similar to linen.[150] Today's hemp fabrics are typically blends of hemp with other natural fibers, such as flax, cotton, or silk. Hemp crops benefit the soil, enriching it for subsequent crops. Because hemp is generally not affected by disease, it rarely needs pesticides. Hemp cultivation uses one-half the land that cotton does and only one-third the water.[151]

The United States used hemp extensively during World War I for uniforms, canvas, and rope. During World War II, the government produced a short film titled *Hemp for Victory*, promoting hemp as a crop that would help win the war.

Hemp fabric is known for its ability to wick moisture away from the body and dry quickly, leaving the individual warmer in a cold climate and cooler in a hot or humid environment. The fabric is naturally antibacterial and resists odors. Hemp fabric is preferred by many with chemical sensitivities.

CLEAN AIR

The largest area of a home, and the one that has the most immediate impact on health, is one that is commonly overlooked when cleaning: the air. We may consider the quality of outdoor air if we live in a heavily populated area or one with severe air pollution. Indoor air, however, which takes up more space by volume than any other area of a home and is taken into the body with every breath, is often left out of discussions about cleaning or greening the home.

Historically, indoor air quality has been a concern. Enclosed spaces led to poor air circulation, less fresh air, and, ultimately, health problems. Better sanitation and hygiene practices lessened these effects, but the current reliance on chemicals and synthetic products has led to other health issues related to indoor air quality.

Air pollution is the largest single environmental cause of illness in children worldwide and is the second leading cause of illness after poor sanitation and a lack of clean water. Indoor air pollution is ranked as one of the world's largest health risks.[152] In industrialized countries, where the average person spends between 80 and 90 percent of his or her time indoors,[153] this is a major health concern.

AIR POLLUTION FROM HOUSEHOLD CLEANERS

Indoor air pollutants, which include everything from paints, varnishes, and adhesives to clothing and building materials, are more hazardous than we might realize. In some places, indoor air quality may be 12–14 times more polluted than the air outdoors.[154] In addition to household items, which may be off-gassing, the very products we use to clean our homes may be causing our air to become toxic. Commercial cleaning products often contain volatile organic compounds (VOCs), including benzene, xylene, hexane, heptane, octane, decane, trichloroethylene, and methylene chloride. Exposure to VOCs is linked to increases in acute illnesses, such as respiratory, neurological, and reproductive disorders, along with various cancers.[155] As we spray and use chemicals during cleaning, fine particles are released into the air. We then breathe the air, including the chemicals, exposing our lungs and bodies directly to various toxic substances.

COMMERCIAL AIR FRESHENERS

"The best way to address residential indoor air quality pollution usually is to control or eliminate the source of the pollutants and to ventilate the home with clean outdoor air." — U.S. Environmental Protection Agency

Some of the most toxic household chemicals are found in commercial air fresheners. Most air fresheners work not by eliminating odors through cleaning, ventilation (dilution), or absorption, but by adding chemicals to the air. Commercial air fresheners work in two ways. Some air fresheners coat a consumer's nasal passageways with chemicals that interfere with nerve endings, lessening the perception of bad odors. Other air fresheners mask odors with a stronger fragrance, which overpowers a consumer's perception of offending smells. The original odors still exist; the consumer just doesn't notice them as much.

Commercial air fresheners often include volatile organic compounds (VOCs), chemicals classified as carcinogenic and neurotoxic; many of them are also known endocrine disruptors.[156] Due to their particulate nature, air fresheners are more easily inhaled into the lungs and delivered into the bloodstream than other household products.

Consumers already face respiratory risks associated with inhaling the chemicals themselves. But risks from interactions between chemicals and the air we breathe may be even worse. Researchers have discovered a link between commercial air freshener use and cardiovascular health. Even limited exposure to commercial air fresheners resulted in adverse heart effects, and more frequent exposures resulted in less cardiac resilience when dealing with stress.[157]

DEALING WITH ODORS

We can deal with odors in our homes in three basic ways: (1) We can prevent odors by removing items or cleaning them before they become a problem; (2) odors can be absorbed or dispersed in order to keep them from being bothersome; or (3) we can cover odors with more attractive scents.

PREVENTING ODORS

Most air fresheners are used in an effort to deal with odors. Rather than look for ways to merely cover up odors, look at preventing or minimizing potential odors before they become an issue.

> Eliminate odors easily by making sure that garbage and compost are in containers and removing the contents of the containers frequently.

Household odors typically originate in the kitchen, the bathroom, or the laundry room. By keeping up with cleaning, and by putting soiled items in containers and removing the soiled items often, we can greatly minimize odor issues.

Garbage and Compost Bins

- Keep garbage in closed containers, such as in lidded trash cans or in cabinets.

- Take garbage out frequently to prevent a buildup of odors from rotting items or odors associated with bacteria or mold.

- Indoor compost bins should be emptied frequently. Removing rotting items found in garbage and compost bins will also help prevent insect issues.

- Wash trash and compost receptacles frequently to remove any offending residue.

- Use a homemade deodorizer to absorb smells associated with trash.

Kitchen Odors

- Use exhaust fans to remove cooking odors and smoke from the kitchen.

- Cook with an appropriate amount of heat to avoid smoke.

- Clean up spills on cooking surfaces.

- While cooking, use items that absorb smells, such as white distilled vinegar in a bowl, slices of bread on a plate, or vanilla extract in a cup.

- Run citrus peels through the garbage disposal to freshen the sink.

- Clean out the refrigerator and pantry.

- Rotate foods in the fridge and pantry so that you use them before they go bad.

- Dispose of any foods that are expired or rotting.

- Clean food spills immediately to prevent them from becoming an issue.

- Use a homemade deodorizer to absorb normal food odors.

Soiled Laundry

- Keep soiled laundry in covered hampers to contain odors.

- Wash frequently to eliminate odors and prevent issues with bacteria and mold growth.

Bathroom Odors

- Use bathroom exhaust fans to remove excess moisture and to prevent mold or mildew.

- Launder towels frequently to prevent any dirt- or moisture-related issues.

- Clean bathrooms frequently to remove dirt and bacteria.

- Use homemade deodorizers to absorb odors.

ABSORBING ODORS

> Avoid adding toxins to your home from commercial air fresheners. Naturally absorb odors with common ingredients, such as baking soda, vinegar, or coffee.

Items that absorb odors are often easy to make and may reduce offensive smells lingering in the home.

BAKING SODA AIR DEODORIZER

1 small box baking soda
20–30 drops of your favorite essential oil

Mix baking soda with essential oil. For a clean smell, consider using citrus oils, such as lemon, orange, lime, or grapefruit. Place in an open container near odorous areas. Alternatively, sprinkle the scented baking soda on any material, such as the carpet, a mattress, upholstery, or shoes, that needs to be deodorized.

VINEGAR DEODORIZER

It may sound counterproductive because of vinegar's strong scent, but leaving a bowl of vinegar out will help neutralize smells.

COFFEE DEODORIZER

Coffee grounds can absorb odors while giving off their own scent. This is popular with coffee lovers and can be augmented by adding extracts or essential oils to the coffee grounds.

Make your own deodorizing discs to place in those suspect areas.

DEODORIZING DISCS

Deodorizing discs are great in closed containers, such as diaper pails, trash cans, or cabinets. Replace the discs when they are no longer absorbing odors.

10–20 drops essential oils of your choice
1–2 cups water
2 cups baking soda
Silicone molds or muffin baking cups

Mix essential oils in approximately 1 cup of water. Add the baking soda and mix well. Add more water until you have a thick paste. Transfer the mixture to your silicone molds or muffin baking cups, and allow to dry for 24–48 hours. When discs are completely dry, remove from molds.

ESSENTIAL OIL GEL DEODORIZER

2 packets gelatin

1 cup water

1 tablespoon vodka

15–20 drops essential oil

Small jar

Mix gelatin packets with water in a saucepan. Bring mixture to a boil while stirring. Allow to cool. Add vodka and essential oil to the mixture, and stir well. Pour mixture into the small jar. As the water evaporates, the scent of the essential oil will disperse throughout the room. The vodka acts as a preservative so that the gelatin does not mold.

DEALING WITH HEAVY ODORS

With heavy odors that are difficult to remove, including cigarette smoke, mold, or mildew, there may be a tendency to turn to toxic chemicals in an attempt to eradicate the smells. This is both unnecessary and harmful, as these chemicals do not remove the offending substances but merely add to the toxic load.

Activated carbon

Activated carbon, also known as activated charcoal, is a form of processed carbon that has an increased surface area for adsorption of odor-causing molecules.[158] It is often produced from coal but can also be made of other sources, such as coconut husk, peat, or wood.

ACTIVATED CARBON DEODORIZER

Activated carbon
Small open pan

Place activated carbon in a small open pan or other container. Allow it to sit and absorb odors.

Note: Activated carbon depletes oxygen levels in the air when wet,[159] so use caution when placing activated carbon in small, wet spaces.

OZONE GENERATORS

Stratospheric ozone, residing at 6–30 miles above the Earth, is naturally occurring and acts as a protective shield against ultraviolet (UV) rays. Most ground-level ozone is created by chemical reactions between nitrogen oxides and VOCs in the presence of sunlight.

Ozone generators use ultraviolet (UV) lamps or electrical discharges to produce ozone, which then reacts with chemical or biological pollutants to neutralize them. Because ozone is a severe mucus membrane and lung irritant,[160] ozone generators should be used only for severe smoke, mold, or mildew issues. Ozone is a powerful oxidizing agent and is therefore toxic,[161, 162] but it is very unstable and breaks down readily into oxygen in the atmosphere. Make sure that people and pets are not near the ozone generator during use, and allow enough time after the generator is turned off for any remaining ozone to decay before people or pets return to the area.[163] Long-term effects from the use of ozone generators are unknown.[164]

AIRING OUT

Open your windows to allow in fresh air, which naturally disperses and removes odors and toxins from your home.

Airing out a home is an important aspect of keeping it clean. Indoor air quality continues to decline in part due to greater use of chemicals and synthetic products. However, energy-efficient homes are also partly to blame, as they block air drafts and prevent cross ventilation of outdoor and indoor air. Open a window on low pollen days — even during winter — to allow fresh air to circulate, thus diluting the concentration of indoor pollution.

PLANTS THAT PURIFY

Plants clean the air. They take in carbon dioxide, converting it to oxygen, which is then released into the atmosphere, making the air more breathable. In the 1980s, researchers at the National Aeronautics and Space Administration (NASA) were looking at ways to purify air in confined spaces for the International Space Station.[165] They needed a self-sustaining, contained system to clean the air in the space station during extended periods while researchers lived there, so they researched plants for prolonged use on the space station. Since then, further research by NASA and other sources has revealed how plants clean the air of impurities and pollutants, including many highly toxic substances.[166]

According to NASA, twenty good-size houseplants can purify the air in an average 2,000 square-foot home. This equals one plant per 100 square feet of indoor space.

Plants are able to absorb impurities from the air through their leaves during respiration and convert those substances into products they can use, such as organic acids, sugars, and amino acids.[167] Anything that the plant is unable to break down is moved along through the plant's structure to the root system, at which point the toxic chemicals are further broken down by microbes in the rhizosphere (the soil surrounding the plant's roots) into food and energy for both plant and microbes. These microbes rapidly adapt, becoming more efficient at breaking down chemicals over time.

Homes with more plants have more negative ions, which are created when outside energy is placed on a molecule, displacing electrons to other molecules nearby. Plant transpiration (in which watery vapor moves out of a plant via a membrane or pores) is one way that this electron displacement occurs.[168] Because synthetic materials — especially those that off-gas in homes, such as building materials — remove negative ions from homes, sufficient plants are needed to counteract those effects.

Negative ions are needed for overall good health and a sense of well-being. This is one reason that houseplants are reported to lessen stress in work environments and increase productivity.[169] Higher levels of negative ions may also lead to reduced dust levels, a common cause of respiratory issues. In a study conducted by the Washington State University, researchers found that plants could reduce dust levels by 20 percent.[170] In addition, plant-filled rooms in which the soil is not exposed have 50–60 percent fewer airborne microbes, such as mold spores.[171]

Top Houseplants for Cleaner Air

■ Add houseplants to your home to naturally purify the air. Spider plants and golden pothos are both easy to care for and naturally remove toxins, such as formaldehyde.

Houseplants are highly effective air purifiers. Including plants in your home or office will quite literally result in a greener space. Because they contribute a better oxygen concentration, add to a pleasant atmosphere, and remove toxic chemicals, plants are important to your health.

Note that although these plants are air purifiers and have been used in households for centuries, not all plants are completely benign. Many common household plants have developed a natural defensive mechanism against being eaten and may contain substances that are toxic to some degree when ingested. Keep plants out of reach of small children or animals, and read about any plants you take into your home. With proper care, plants will make a wonderfully healthy addition to your abode.

All of the following plants will help rid your home of toxins in the air. If you are confused about what plant to buy, consider looking at ferns, ivies, and spider plants, which are easy to grow and quite effective.

ALOE (*ALOE VERA*): Aloe vera plants are well known for their medicinal qualities. Cut a leaf open and slather some of the gel you find inside on cuts and burns.[172, 173] These plants also remove many of the chemicals commonly found in household cleaners, such as benzene and formaldehyde. Aloe vera plants are easy to grow in sunny locations and require minimal watering.

ARECA PALM (*DYPSIS LUTESCENS*): This plant should be allowed to dry slightly between waterings, as it does poorly with overwatering. The areca palm is one of the top air-purifying plants, ranking high for overall purification and removal of xylene and toluene.[174]

AZALEA (*RHODODENDRON SIMSII*): Azaleas are ornamental flowering shrubs with large blossoms. They do well in cooler temperatures with bright light. Azaleas are particularly adept at removing chemicals off-gassed from synthetic sources.[175]

BAMBOO PALM (*CHAMAEDOREA SEFRITZII*): Bamboo palms are small palms with tall, canelike stems that end in pinnate leaves (leaves that resemble feathers).[176] Besides removing formaldehyde from the air, bamboo palms are natural humidifiers.

BOSTON FERN OR SWORD FERN (*NEPHROLEPIS EXALTATA*): Boston ferns are tropical plants often found in hanging pots, which allow their natural sweeping fronds to hang over the sides. Because they are tropical, they prefer a lot of indirect light. The soil must be kept damp, and misting may be necessary for a healthy plant. Boston ferns are considered the number one plant for removing formaldehyde from your home and are efficient at removing mold spores from the air. If your bathroom has a lot of indirect natural light, consider keeping a Boston fern in it.

CHINESE EVERGREEN (*AGLAONEMA CRISPUM "DEBORAH"*): Chinese evergreen is one of the most popular ornamental plants because it is easy to grow. It flourishes in low light. It is a top-rated plant for filtering toxins from the air.[177]

CHRYSANTHEMUM (*CHRYSANTHEMUM MORIFOLIUM*): Perennial flowering plants, chrysanthemums are prized for their lovely blooms but are also used in cooking and as a natural insecticide. For this reason, many gardeners plant the hardy version in garden areas and in pots around outdoor living areas. NASA lists mums in their Clean Air Study for use in indoor air purification.[178]

ENGLISH IVY (*HEDERA HELIX*): English ivy is cultivated as an ornamental plant. Indoors, it filters both benzene and formaldehyde. The plant thrives in most household environments but prefers bright, filtered light.

GERBER DAISY (*GERBERA JAMESONII*): Used in households and gardens for their beauty, these plants also remove benzene and trichloroethylene from the air. Unlike many plants, Gerber daisies continue to release oxygen during the night, making them wonderful additions to bedrooms.

GOLDEN POTHOS OR DEVIL'S IVY (*EPIPREMNUM AUREUM*): These do well in a variety of lighting settings and tolerate various water conditions, making them excellent plants for those who have difficulty keeping plants alive. Golden pothos are efficient in removing formaldehyde, xylene, and benzene.[179, 180]

HEART LEAF PHILODENDRON (*PHILODENDRON OXYCARDIUM*): The philodendron is native to tropical jungles and tends to prefer medium light and indoor temperatures, making it perfect for many areas of a home.[181] The trailing vines of this plant should be kept high, as this plant is a common culprit in ingestion toxicity in small animals and children. The plant is excellent at filtering toxic chemicals from the air.

MOTHER-IN-LAW'S TONGUE OR SNAKE PLANT (*SANSEVIERIA TRIFASCIATA "LAURENTII"*): Although this plant prefers moderate to bright light, it does well under a multitude of conditions. The snake plant removes air pollutants, including formaldehyde and nitrogen oxides.[182]

PEACE LILY OR SPATH LILY (*SPATHIPHYLLUM*): These plants enjoy a damp environment and remove mold spores from the air. They also remove numerous contaminants from the air, including trichloroethylene, formaldehyde, and benzene.[183]

RED-EDGED DRACAENA (*DRACAENA MARGINATA*) AND WARNECK DRACAENA (*DRACAENA DEREMENSIS "WANECKLII"*): Dracaena are a genus of trees and succulent shrubs. They enjoy low light conditions and are therefore often used in corners and hallways. They require watering only every 7–10 days and are effective at general air filtration.

SPIDER PLANT (*CHLOROPHYTUM COMOSUM*): This plant thrives in indirect natural light and needs only light watering. Ranked by NASA as one of the top three plants for removing formaldehyde, the spider plant is also especially efficient at removing xylene and carbon monoxide from the air.[184]

WEEPING FIG, BENJAMIN'S FIG, OR FICUS TREE (*FICUS BENJAMINA*): The ficus tree's elegant growth and tolerance for a wide range of growing conditions make it a favorite with homeowners, whether experienced or inexperienced with plants. Ficus trees have proven effective at removing formaldehyde from the air.[185]

Herbs That Clean

It's not just houseplants that clean: herbs also purify the air by filtering out toxic chemicals during their normal respiration. Not only are they great at cleaning the air, herbs are attractive, smell good, and have culinary and medicinal uses. When grown indoors, herbs can be harvested year round and used fresh in cooking and herbal remedies. The aromatic nature of herbs adds pleasing scents to the air and can neutralize cooking odors in the kitchen. If you would like to add some multipurpose herbs to help clean the air in your home, check out some of the following:

- Basil (*Ocimum basilicum* L.)
- Bay (*Laurus nobilis*)
- Chervil (*Anthriscus cerefolium*)
- Chives (*Allium schoenoprasum*)
- Dill (*Anethum graveolens*)
- Jasmine (*Jasminum officinale* L.)
- Lavender (*Lavandula spica* L.)
- Marjoram (*Origanum majorana*)
- Mint (*Mentha spicatta* L.)
- Oregano (*Origanum vulgare* ssp. *hirtum*)
- Parsley (*Petroselinum* spp.)
- Rosemary (*Rosmarinus officinalis* L.)
- Sage (*Salvia officinalis*)
- Thyme (*Thymus* spp.)
- Tarragon (*Artemisia dracunculus*)

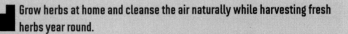

Grow herbs at home and cleanse the air naturally while harvesting fresh herbs year round.

ESSENTIAL OILS

Essential oils are an easy way to add scent to a home. Not only do they have pleasing aromas, but many essential oils also have therapeutic and cleansing properties (page 177). Diffusion is the process by which essential oils are dispersed throughout the air, filling a room or area with natural fragrance.

Tissue Diffusion

Tissue diffusion uses a small piece of material, such as a tissue or a cotton ball, on which a few drops of essential oil are placed. This is an ideal way to scent small, enclosed spaces or to carry a scent with you, but it is not effective for large areas. Try this technique for adding essential oils to cabinets, drawers, or garbage areas. Specific essential oils, such as lavender or cedar, have their own added benefits, such as repelling insects.

Steam Diffusion

Essential oils can be added to heated water for easy steam dispersal. The oils are added directly to a pan of water, which is heated on the stove, or they are added to the well of a humidifier. Oils can be chosen for their therapeutic properties. For instance, eucalyptus and peppermint are often used for respiratory ailments. Relaxing oils may be used to aid sleep. Steam diffusion quickly disperses the scent throughout a room but is not long-lasting.

Candle Diffusion

Candle diffusion requires an all-natural soy or beeswax candle. Avoid using candles made from petroleum products such as paraffin. Light the candle and allow a small pool of wax to melt. Extinguish the candle and add a drop of essential oil to the pooled wax before relighting the candle. A modern version of this method uses a standard light bulb and a metal or ceramic ring with a few drops of essential oil placed on it. As the light bulb heats the ring, the essential oil is dispersed into the air. Take special care when using candle diffusion, as essential oils are highly flammable.

Fan Diffusion

Fan diffusers can be purchased in a multitude of sizes and shapes. After essential oils are placed on an absorbent pad or in a tray, electricity is used to power a fan, which disperses the scent into the air. Many fan diffusers are noisy, or they might require that you purchase replacement pads. A similar technique is to place a small bowl of essential oils in front of a fan that you are using to cool your home.

Electric Diffusers

Much like candle diffusion, electric diffusers use electricity to heat an essential oil. Depending on the individual unit, they can scent a large area. If the essential oil is heated for too long, however, certain components of the essential oil may be destroyed, lessening its benefits.

Nebulizer

A nebulizer (or atomizer) breaks essential oils into separate molecules before dispersing them into the air. Smaller particles mean that the oils are more readily absorbed through the lungs, which may be desired if essential oils are being used for their health benefits. Nebulizers are more expensive than the other diffusion methods and may need replacement parts if their glass pieces break. Thick oils can clog some nebulizers.

ESSENTIAL OILS SPRAY

Essential oils may also be used in spray bottles, along with a carrier liquid. You can spray the mixture into the air or on household items to freshen the air. This is a quick method of diffusion but is generally not long-lasting.

2 cups water
3 tablespoons vodka
10 drops essential oil of your choice

Combine all ingredients in a spray bottle. Shake well before use.

UPHOLSTERY SPRAY

1 cup water
1 cup white distilled vinegar
½ teaspoon vegetable glycerin
10 drops essential oil

Combine all ingredients in a small spray bottle. Shake well, spray upholstery, and allow to evaporate. Although these ingredients are not known to discolor fabric, it is prudent to try the spray on an inconspicuous area first.

TOILET PAPER ROLLS

Try adding a few drops of oil to the inside of a toilet paper roll before placing it in the bathroom. The essential oil will waft throughout the room, adding a pleasing scent.

EXTRACTS

Avoid artificially scented cleaning supplies by perfuming your homemade cleaning supplies with your own easy-to-make extracts: soak vanilla, citrus peels, aromatic leaves, or almonds in vodka.

You might think of extracts as being only for cooking, but extracts also can be used to scent the home. These inexpensive air fresheners are easy to make and use. Simmer an extract on the stove, and it will permeate the air with a natural aroma that is safe for humans and pets alike. Alternatively, try placing a bowl of homemade extract in a heavily used area for a subtle, pleasing scent, or add extract to homemade potpourri. As the alcohol evaporates, the scent is carried into the air.

When making extracts, consider preparing large amounts. Not only can you use these around your home, but they are wonderful for cooking. Homemade extracts are generally of better quality than commercially made versions, and because you made them, you know exactly what is in them. Extracts also make beautiful gifts that recipients will appreciate.

VANILLA EXTRACT

2–3 vanilla beans
1 cup vodka
Bottle with lid

Slice vanilla beans lengthwise, leaving the ends intact so that you can more easily remove them later. Pour the vodka into a bottle. (If you are making a large amount of vanilla extract, just add beans to the bottle of vodka.) Immerse the cut vanilla beans in the vodka. Put the lid on the bottle, and place the bottle in a cool, dark area. Let the vanilla sit for six weeks or more for a flavorful extract. The longer you leave the vanilla beans in the vodka, the richer your vanilla extract will be.

ORANGE EXTRACT

2 oranges
1 cup vodka
Bottle with lid

Use a vegetable peeler or zester to remove the peel from two oranges. Place the orange peel, minus any pith, and the vodka in a bottle, and put the lid on the bottle. Place the bottle in a cool, dark location, and let the concoction sit for at least six weeks. For a more flavorful and aromatic extract, allow the peel to remain in the vodka longer.

LEMON EXTRACT

4 lemons
1 cup vodka
Bottle with lid

Use a vegetable peeler or zester to remove the peel from four lemons. Place the lemon peel, minus any pith, and the vodka in a bottle, and put the lid on the bottle. Place the bottle in a cool, dark location, and let the concoction sit for at least six weeks. For a more flavorful and aromatic extract, allow the peel to remain in the vodka longer.

GRAPEFRUIT EXTRACT

2 grapefruits
1 cup vodka
Bottle with lid

Use a vegetable peeler or zester to remove the peel from two grapefruits. Place the grapefruit peel, minus any pith, and the vodka in a bottle, and put the lid on the bottle. Place the bottle in a cool, dark location, and let the concoction sit for at least six weeks. For a more flavorful and aromatic extract, allow the peel to remain in the vodka longer.

ALMOND EXTRACT

12 raw almonds
1 cup vodka
Bottle with lid

Pulverize the almonds with a food processor or blender until they resemble a coarse salt or flour. Place the almond meal and vodka in a bottle, and put the lid on the bottle. Store in a cool, dark area for at least several months. For a more flavorful and aromatic extract, allow the almond meal to remain in the vodka longer. If using for consumption, you may want to strain the almonds out of the extract to eliminate graininess.

PEPPERMINT EXTRACT

½ cup fresh peppermint leaves
Bottle with lid
1 cup vodka

Place the peppermint leaves in the bottle, and pour the vodka over them. Put the lid on the bottle, and place the bottle in a cool, dark place for several months for a lovely peppermint extract.

CINNAMON EXTRACT

4 cinnamon sticks
1 cup vodka
Bottle with lid

Place the cinnamon sticks and vodka in the bottle. Put the lid on the bottle. Let the bottle sit in a cool, dark place for at least several months before removing the cinnamon sticks. For a more concentrated cinnamon extract, allow the cinnamon sticks to stay in the extract longer.

Potpourri and Sachets

Sold in many stores, commercial potpourri, with its synthetic and often toxic fragrances, only vaguely resembles the homemade version. Making your own potpourri, a mixture of dried ingredients, is easy. Stronger scents can be created by adding essential oils or extracts to the dried ingredients. Homemade potpourris are meant to lend a subtle scent to an area without being overpowering. Customized scents can be created by varying the ingredients.

There are two main ways to use potpourri. The first is to place the potpourri in a small bowl or jar. You may choose to close the jar most of the time, opening it only for short periods to lightly scent the room. Or you may prefer to leave the potpourri out all of the time to permeate the air.

Alternatively, place the dried items in water and heat the mixture to create a stronger scent. This can be done in a small pot using electric or candle heat, or in a pot on the stove. Scents can be varied by season or mood to create a pleasing aromatic atmosphere.

When drying plant materials or potpourri, items should be allowed to air dry. Items that are overdried will lose their natural essential oils, rendering them ineffective.

EASY POTPOURRI

6 drops essential oils
1 cup dried plant matter
Spices (optional)

Add essential oils to the dried plant matter and any spices you desire. Gently mix. Store in an airtight container for several weeks to allow the essential oils to permeate the dried plant matter, stirring occasionally. Unused potpourri should be placed in a cool, dark place.

Dried Potpourri Ingredients

Experiment with different concoctions to find the scents you prefer. Dried ingredients for potpourri and sachets can be purchased from your local natural foods store or can be made from gathered plants. To dry ingredients at home, hang plants to dry, lay small items out where they can have a good air flow, or use a dehydrator. Avoid buying dried items at craft stores, as these generally have synthetic fragrances and toxic chemicals added to them. Try drying various flowers and herbs to make a scent unique to your home.

Some popular ingredients for use in potpourri and sachets:

- Allspice
- Cedar shavings or essential oil
- Cinnamon sticks, extract, or essential oil
- Cloves
- Cypress shavings or essential oil
- Fennel seeds
- Jasmine flowers or essential oil
- Jujube flowers or essential oil
- Juniper shavings or essential oil
- Lavender leaves, flowers, and essential oil
- Lemon balm leaves, flowers, and essential oil
- Lemon peel or essential oil
- Marjoram leaves and flowers
- Mint leaves, flowers, or essential oil
- Orange peel or essential oil
- Pine shavings, cones, or essential oil
- Rose flowers, hips, or essential oil
- Rosemary leaves or flowers

SACHETS

Potpourri sachets are a fantastic way to refresh your drawers or closets. They can also be used in vehicles, sports bags, or shoes. Sachets made with potpourri can be customized for the area. For example, sachets placed near stored clothing may also be used to help repel insects. Sachets placed in sports bags may need to be antibacterial.

6–12 drops essential oils
1 cup dried plant matter
Spices (optional)
All-natural cloth bags, such as cotton or silk

Add drops of essential oil to the dried plant matter and any spices you desire. Gently mix. Store in an airtight container for several weeks to allow the essential oils to permeate the dried plant matter, stirring occasionally. Transfer the mixture to all-natural cloth bags, where the scent will be able to permeate the fabric, scenting your home without adding unwanted chemicals.

ORANGE POMANDERS

Pomanders were historically used in homes to scent the air during winter, when individuals were less likely to bathe or wash clothing frequently. They are attractive, easy to make, and can be placed wherever needed. Pomanders are a popular decoration during winter holidays.

Toothpick
1 orange
Whole cloves

Poke holes through the skin of the orange with the toothpick to enable easy insertion of the cloves. This can be done randomly or in a design. Push the smaller end of a clove into a hole until the large end is next to the peel. Continue until you have filled all of the holes you made. As the moisture in the orange evaporates through the cloves, the air will be filled with a pleasant aroma.

Pomanders are typically made with oranges, but other citrus fruits can be used. Try experimenting with different fruits for a wonderful pomander display that scents your home.

OUTDOOR LIVING

Spending time outside isn't important only for "outdoorsy" people. Getting out of doors is an essential aspect of staying healthy. Outside we can breathe fresh air and get away from off-gassing inside our homes. Going outside also encourages us to be more active.

The outdoors are important for our health, but there are risks outside to consider, too. Luckily, there are some simple things you can do to make your time outside a greener, healthier experience.

GRILLING: LOWERING YOUR IMPACT AND EXPOSURE

Grilling is a summer pastime. Take your cooking outdoors so that you can enjoy the weather and eat some tasty food without heating up your kitchen and home. When it comes to your environmental impact and your health, there are a few things to take into consideration.

GAS OR CHARCOAL?

Gas grills, whether using natural gas or propane gas, are actually better for the environment and better for you than charcoal grills. They use less energy and produce fewer pollutants, meaning fewer risks for you and your family. Scientists in the United Kingdom conducted an Environmental Impact Assessment Review to compare gas versus charcoal grills. They found that a charcoal grill will emit 2,200 pounds of carbon dioxide over its lifetime.[186] That is almost three times the carbon footprint of a natural gas grill!

Charcoal grills also produce hydrocarbons, along with sooty particles, which pollute the air and can aggravate heart and lung problems. Both lump charcoal and charcoal briquettes create air pollution. Lump charcoal, a form of wood that is charred to add flavor, contributes to deforestation and adds to the greenhouse gases in the atmosphere. Although charcoal briquettes do make use of waste wood by incorporating sawdust, many brands also contain other fillers, such as coal dust, starch, sodium nitrate, limestone, or borax.

Charcoal can also harm your health by adding carcinogens to your food. According to the American Cancer Society, grilling meat over charcoal can result in two kinds of potentially carcinogenic compounds. The hotter and longer the meat cooks, the more carcinogenic compounds are formed.[187]

In Canada, charcoal is now restricted under its Hazardous Products Act. Charcoal briquettes advertised in, imported into, or sold in Canada must have a label warning of potential hazards associated with the product.

Lower-Impact Charcoal

Gas grills do tend to be more expensive than their charcoal counterparts. If a gas grill is out of your price range, or if you already have a charcoal grill, look for natural charcoal brands to minimize health risks. Some brands are made of 100 percent wood or coconut shells. These natural brands should not contain any coal, oil, limestone, starch, sawdust, or petroleum products. Look for a product that is 100 percent natural with no fillers. Also, look for lump charcoal made from certified sustainably harvested wood.

Looking to add a little flavor to your gas grill with soaked wood chips? Again, look for sustainably harvested wood. If you can't find any that are certified sustainable, look for hardwoods, such as hickory or mesquite.

GETTING STARTED: LIGHTING YOUR FIRE

What about getting started? Is it safe to use lighter fluid? In a word, no. Some areas of the United States have banned the use of lighter fluid to start backyard grills due to the volatile organic compounds (VOCs) released into the ozone.[188] Anything you burn is being released into the air and directly into your food. Self-lighting charcoal is no better, as it also contains chemicals. If your fire-starting skills are lacking, there are all-natural fire starters available, such as fatwood, which is wood that has been soaked in the tree's natural resin. Or try a charcoal chimney or an electric charcoal starter.

CLEANING YOUR GRILL WITHOUT CHEMICALS

Food or grease left on the grill can shorten its life. Replacing your grill every few years due to neglect is neither economical nor environmentally friendly. Leftover food from the last time you grilled doesn't sound appetizing, and if not taken care of, it can make you sick. You don't need to use toxic commercial cleaners on your food surfaces, though. There are some easy ways to safely clean your grill.

> Avoid the toxic chemicals found in grill cleaners by using vinegar and baking soda to clean your grill.

ALUMINUM FOIL

After the last item is taken off the grill, don't just turn off the grill. Lay some aluminum foil on it, and let the heat continue to burn off leftover foods. The next time you grill, crumple the foil and easily scrub off the burned and sooty remains. Start up the grill the way you normally would.

BAKING SODA GRILL PASTE

¼ cup baking soda
¼ cup water

Make a paste out of the baking soda and water. Apply to the grill and allow to dry for approximately 15 minutes. Wipe the area with a clean, dry cloth, and then heat for 15 minutes to burn off any residue before placing food on the grill.

BAKING SODA AND VINEGAR SPRAY

Baking soda
Spray bottle of white distilled vinegar

Burn off as much food residue as possible after cooking. Allow the grill to cool before sprinkling with baking soda. Using a spray bottle filled with white distilled vinegar, spray the entire area. The baking soda and vinegar will react, producing bubbles that will work to remove any remaining grease. Continue spraying to keep the baking soda damp and the reaction going, then wipe with a wet rag.

STAINLESS STEEL GRILL POLISH

1 cup baking soda
¼ cup borax
½ cup lemon juice
Club soda as needed to make a paste

Mix dry ingredients. Add the lemon juice. The mixture will begin to form a dry paste. Add club soda until the paste is the consistency you like. Apply the paste to the stainless steel grill, and allow paste to sit for a few minutes. Wipe off with a wet cloth, and rinse away any residue. To prevent spotting on stainless steel, use a dry cloth to wipe dry.

PATIOS AND DECKS

Dirt on outdoor furniture may seem heavy-duty, but you don't need harsh chemicals to get it clean. Here are a few recipes you can use to clean your patio and your patio furniture.

PLASTIC FURNITURE

Many of the solutions from the "All-Purpose Cleaners" section (page 56) are perfect for cleaning your plastic outdoor furniture.

Using a scrub brush or sponge, try cleaning your furniture with the Amazing All-Purpose Cleaner (page 57). If your plastic furniture is white and in need of brightening, try the Better-Than-Bleach Disinfectant (page 60). If mildew is still present after scrubbing with an all-purpose cleaner, use the Mold and Mildew Cleaner (page 108).

OUTDOOR CUSHIONS

TOUGH STAIN CLEANER*

Washing soda
Water

Make a paste out of washing soda and water. Using a wet cloth or sponge, scrub the stains with the paste. Rinse well.

* For tough stains on white fabric, try spraying the fabric with hydrogen peroxide. Let it sit for a few minutes before using the Tough Stain Cleaner or the Light Cleaner.

LIGHT CLEANER*

2 cups water
½ cup castile soap

Mix castile soap with water for a soapy solution. Dab or spray onto soiled surfaces. Scrub with a wet cloth or sponge.

* For tough stains on white fabric, try spraying the fabric with hydrogen peroxide. Let it sit for a few minutes before using the Tough Stain Cleaner or the Light Cleaner.

CANVAS CHAIRS

Removable canvas chair covers can be tossed in your washing machine with some soap. Remove the clean, wet canvas from the washing machine, and put it back on the chair to dry in order to maintain its shape. Canvas that cannot be removed should be cleaned in the same way as outdoor cushions.

WOOD

WOOD PATIO FURNITURE CLEANER*

2 cups water
2 tablespoons castile soap

Mix water and castile soap in a spray bottle. Lightly spray wood with the mixture, and wipe off with a soft cloth; use a soft-bristle brush for crevices or wicker, as needed.

* To add a little bit of color and shine, finish with a soft cloth that is sprinkled lightly with a vegetable- or plant-based oil. Olive oil, jojoba oil, or sweet almond oil are all good choices.

WROUGHT IRON

To remove rust without the aid of chemicals, you'll want to use a power tool, such as an electric sander, to scour the rust from iron furniture. Prevent wrought iron furniture from rusting by keeping it dry. Move it inside or cover it with a waterproof fabric when not in use. If your furniture needs a protective coating, try rubbing some solid beeswax onto the dry metal surface. This will prevent moisture from reaching the metal, but may need to be redone several times a season as temperatures heat up.

RUST STAINS ON CONCRETE

CONCRETE RUST STAIN REMOVER

White distilled vinegar

Pour vinegar onto rust stain. Let sit for several minutes. Scrub stain with stiff brush.

DECK AND PATIO MOSS CLEANER

White distilled vinegar

If moss is plaguing your outdoor living spaces, try a little white distilled vinegar. Just pour straight vinegar on the area, and go do something else for an hour or two. When you come back, rinse the area with a hose. Repeat as needed.

POWER WASHERS

When it comes to outdoor cleaning, power washers are your friend. Whether you are cleaning a deck or patio, or even a fence or house, power washers make the job easier.

Remember to go slowly. It took years to build up the dirt and grime. You won't clean an entire deck in 30 minutes. Move steadily, and remember that power washers have some kick to them. If you stay in one spot, you may find yourself with grooves cut into the wood from the high pressure.

VEHICLES

You may have heard that you shouldn't wash your vehicle at home. This is true if you are using commercial products made from petroleum distillates, silicon, kerosene, or mineral spirits. The chemicals in these products will run off with the water and will find their way into storm drains and sources of fresh water.

That doesn't mean that you can't safely wash your car at home. You just need to use environmentally safe products and consider your options. Instead of using chemicals to get your car sparkling, try using soap and water.

Tips for getting your car clean:

- Wet the car with a hose.

- Consider parking on the grass. It sounds odd, but parking on the grass rather than on concrete will help filter runoff water before it makes it back to the water table.

- Ditch the paper towels. Instead, grab old towels or T-shirts.

- Don't use chemicals. Use all-natural cleaning products to get your car pristine.

SOAPY CAR WASH

¼ cup castile soap
1 gallon hot water

Add the castile soap to a bucket of hot water. Use the solution to scrub your car. Rinse with fresh water so that a film isn't left behind.

GRIMY-MAT CLEANER

1 part washing soda
1 part salt
1 part water

Mix all ingredients to form a paste. Apply to soiled areas and allow to sit for a few minutes—longer for heavily soiled areas. Using a scrub brush, scrub the mats and rinse as needed. Allow mats to completely dry before putting them back in the vehicle.

CAR WINDOWS: Try your favorite glass-cleaning solution from our glass cleaners section (page 63) to get your vehicle windows sparkling. To prevent spotting, dry the windows with a soft cloth or old newspaper.

INTERIOR SPRAY: Use one of our all-purpose cleaners (page 56) to clean your vehicle's interior surfaces. If using a solution with an acid, such as vinegar or lemon juice, avoid any waxed areas.

JOJOBA CAR CONDITIONER: The interior of your car receives more abuse than you might realize, what with exposure to sunlight and extreme temperatures. Treat it right with a little Jojoba Car Conditioner. Apply jojoba oil with a soft cloth to condition wood, vinyl, and leather. Apply in long strokes and buff with a clean cloth.

WHEEL CLEANER: Wheels and tires get the brunt of the dirt. Make up some Grimy-Mat Cleaner, and use it to help take care of rust and grease. Do not use on painted surfaces.

FROST-FREE WINDSHIELD-WIPER FLUID: Avoid toxic chemicals and mix up some of your own Frost-Free Windshield-Wiper Fluid. Simply add a 50:50 mixture of vinegar and water to the reservoir. This will help keep your windows clean while also preventing frost.

BUGGING OUT

Spending time outside is no fun if you are being eaten alive by insects! Their bites itch, and some insects, such as mosquitoes[189] and ticks,[190, 191] can carry disease. Those with allergies to the insects find themselves even worse off. It's enough to make most people head back indoors, eschewing their time outside for a bug-free evening. Insects may represent 80 percent of the world's species[250] but there are some easy things we can do to repel insects by making us, and our yards, less desirable.

It is estimated that for every human alive on the planet, there are 200 million insects alive.

For most families, an easy solution is to use bug spray. However, commercial versions are not as safe as manufacturers have us believe. In fact, they contain toxic chemicals that should never be on our skin.

THE DANGERS OF DEET

DEET (chemical name, N,N-diethyl-meta-toluamide) is the main ingredient in the majority of commercially sold insect repellents in the United States. DEET repels insects by masking a person's natural odor.[192] Unfortunately, DEET (which ranges in concentrations in commercial insect repellents from 10 percent to 100 percent) can cause numerous adverse health effects, including "neurological effects (seizures, encephalopathy, tremor, slurred speech, coma, and rarely, death) in children and adults."[193, 194, 195, 196, 197, 198, 199, 200, 201] These effects are most often associated with ingestion or skin applications. In 2002, Canada banned the use of products containing DEET in concentrations greater than 30 percent, but products sold in the United States have concentrations as great as 100 percent.[202, 203]

DEET is absorbed through intact skin.[204] This is of particular concern with children, whose thinner skin, along with a greater skin-to-body-volume ratio, absorbs DEET at a faster rate. Research has also shown that DEET can be toxic in combination with other chemicals, including permethrin, another chemical commonly found in commercial insect repellents.[205, 206]

NATURAL BUG REPELLENT

When it comes to applying insect repellent, you aren't limited to toxic chemicals. New research using naturally occurring compounds has shown that their efficacy is equal to chemicals such as DEET, but without the toxic side effects. You may need to reapply the natural repellents more frequently than the commercial versions, but that is a negligible concern when compared with the risks of putting toxic chemicals on your skin. If you would like to make your own insect repellent from essential oils, try these all-natural recipes.

 Don't spray your family with toxic bug sprays. Make your own all-natural bug spray using essential oils, such as lemongrass, citronella, lavender, or lemon eucalyptus. Add water for a spray or coconut oil for a rub.

ESSENTIAL INSECT REPELLENT SPRAY

25 drops essential oils (try lemongrass, citronella, lavender, lemon eucalyptus, and/or rose geranium for their insect-repelling qualities)
¼ cup water (alternatively, use vodka for a quickly evaporating spray)

Mix essential oils and water in a spray bottle. Shake well before each use. Spray over body before heading outside.

ESSENTIAL COCONUT INSECT REPELLENT

10–25 drops essential oil (try lemongrass, chamomile, citronella, lavender, and/or rose geranium—the last is especially good for ticks)
¼ cup melted coconut oil

Add the essential oil to the melted coconut oil. Mix well and store in a small container. The next time you head outside, rub some of the Essential Coconut Insect Repellent on your skin. The essential oils will help repel insects while the coconut oil will nourish your skin and provide minimal UV protection.

INSECT REPELLENT SOAP

Essential oil of choice (try lemongrass, citronella, lavender, lemon eucalyptus, rose geranium, and/or thyme for their insect-repelling qualities)
Castile soap

You can easily add some insect-repelling qualities to your hand soap. Simply add 10–15 drops of essential oil per ounce of castile soap. Wash before and after going outside to minimize insect annoyances.

Also, try adding insect-repelling essential oils to the Foaming Soap Solution recipe (page 110).

MINTY BUG SPRAY

25 drops total of the following essential oils: catnip, spearmint, and/or peppermint
¼ cup water

Mix the essential oils and water in a spray bottle. Shake well before using.

CITRUS BUG SPRAY

25 drops total of the following essential oils: lemongrass, citronella, lemon eucalyptus
¼ cup water

Mix the essential oils and water in a spray bottle. Shake well before using.

WOODSY BUG SPRAY

25 drops total of the following essential oils: cedar, neem
¼ cup water

Mix the essential oils and water in a spray bottle. Shake well before using.

CALMING BUG SPRAY

25 drops total of the following essential oils: thyme, lavender, chamomile, eucalyptus
¼ cup water

Mix the essential oils and water in a spray bottle. Shake well before using.

CUSTOM BUG SPRAY

Are you ready for a bug spray that is custom made for you and your family?

1 part essential oils of your choice (thyme, lemongrass, lavender, peppermint, lemon eucalyptus, citronella, catnip, spearmint)
3 parts carrier (water, vodka, or witch hazel)

Combine your favorite insect-repelling essential oils, and add to a carrier base. Pour in a spray bottle and shake well before use.

CUSTOM BUG RUB

This rub works very well when you or a family member needs insect protection near water, such as when playing in sprinklers or when hanging out near a lake.

1–2 ounces total essential oils of your choice (thyme, lemongrass, lavender, peppermint, lemon eucalyptus, citronella, catnip, spearmint)
8 ounces oil total (sweet almond, olive, coconut, jojoba, and so on)

Combine your favorite insect-repelling essential oils, and add to the carrier oil. Mix well. Store in a glass jar with a tight-fitting lid.

INSECT-REPELLING STREAMERS

Essential oils of your choice
Strips of ribbon or other fabric

Add some essential oil to strips of fabric and hang the strips around your deck or patio. The essential oils will help repel insects and the fabric will give a festive feel. Refresh as needed.

INSECT-REPELLING BRACELETS OR ANKLETS

Essential oils of your choice
Cotton bracelet or anklet

Add essential oils to a cotton bracelet or anklet and allow to dry. Wear when out hiking or playing. Refresh the oil as needed. Similarly, you can use a cotton collar for your dog. Just check with a trusted resource on which essential oils are safe for your pet.

Neem Oil

Neem oil is a natural insecticide made from the seed kernels of the neem tree, *Azadirachta indica*.[207, 208, 209] A mixture of 2 percent neem oil in coconut oil has been shown to give adequate (some say complete) protection against mosquitoes for as long as 12 hours.[210, 211, 212, 213]

Instead of masking odor or repelling insects, the components of neem essential oil act like the hormones produced by various insects. Neem oil enters the insect's system, blocking its hormones and interfering with its normal habits, such as eating, mating, and laying eggs; consequently, the insect population drops. It appears that neem oil does not affect beneficial insects; instead, it targets the hormone receptors of only chewing and biting insects. As neem oil works like a hormone in insects, only a very small amount is needed for use in insect repellents. Topical application of diluted neem oil has shown no adverse effects, but some studies question consumption of the oil[214] via food crops sprayed with neem oil for insect prevention.

REPELLING INSECTS IN YOUR HOUSE

From ants to fleas and from spiders to flies, here are some natural ways to keep critters where they belong — outside. In addition to the recipes below, you can also repel ants and cockroaches by sprinkling diatomaceous earth wherever these two pests are active in your house, at their points of entry to your house, and around the perimeter of your house. (See page 215 for more information on Diatomaceous Earth.)

ANTS

Certain herbal plant extracts have been shown to be 100 percent effective at repelling ants.[215] The herbs give off a strong scent that either repels the ants, kills the fungi the ants feed on, or disrupts the ants' scent trails.

HERBAL PLANT EXTRACT ANT REPELLENT

1 or more of the following plants: cucumber peels, mint leaves, garlic cloves, lemon rind or pulp
Water

Grind one or more of the plant parts into water. Soak cotton balls in the mixture and place along windowsills, doorways, and other points of entrance for ants.

SCATTERED HERBS ANT REPELLENT

If preparing an extract seems daunting, simply scatter or sprinkle some of the following herbs and spices along windowsills, doorways, and other points of entrance for ants:

Cinnamon (sticks or powder)
Bay leaves
Cayenne pepper
Garlic cloves
Paprika
Dried mint leaves (loose or in tea bags)

ESSENTIAL OIL ANT REPELLENT

Essential oils of your choice (some essential oils that repel ants include peppermint, neem, cinnamon, basil, and citrus oils such as lemon and grapefruit)
Cotton balls

Add 10–20 drops total of essential oils to cotton balls, and place at points of entrance for ants. Replace as needed.

COCKROACHES

Sprinkle diatomaceous earth wherever cockroaches are active in your house.

ESSENTIAL OIL COCKROACH REPELLENT

Catnip essential oil[216]
Marjoram essential oil[217]

Add 10–20 drops total of one or both essential oils to cotton balls, and place at points of activity for cockroaches. Replace as needed.

FLEAS

The ideas in this section are more effective at keeping fleas away or managing a small flea population than at dealing with a heavy flea infestation.

CITRUS-MINT CARPET FLEA TREATMENT

3 cups baking soda
15 drops sweet orange essential oil
15 drops citronella essential oil
10 drops peppermint essential oil
10 drops spearmint essential oil
10 drops lemon essential oil

Vacuum carpet before treating. Combine ingredients and sprinkle liberally on carpet. Wait 1–2 hours and vacuum again.

Cedar Oil

Cedar essential oil is a natural flea repellent. Use cedar in any form to naturally drive fleas out of your home. (Note: The ideas in this section are more effective at keeping fleas away or managing a small flea population than at dealing with a heavy flea infestation.)

Place cedar essential oil–soaked cotton balls around your home, particularly in areas that pets have access to. Place cedar essential oil drops on pet bedding. Put cedar blocks under furniture.

FLEA REPELLING SACHETS

Other herbs repel fleas, too. Try making sachets out of the dried leaves of the following plants; wrap the dried leaves in an old cotton hankie. Or soak cotton balls in essential oils of the plants listed below.

Lavender
Mint
Rosemary

Place sachets around pet bedding, furniture, and other problem areas. Refresh as needed.

FLIES

MINTY FLY REPELLENT

Flies are deterred by the smell of mint. Luckily, humans usually enjoy it!

Mint essential oil (peppermint, spearmint, catnip,[218] and so on)
Dried mint leaves

Create a mint-scented sachet by wrapping any natural fabric around mint leaves, or put 10–20 drops of mint essential oil on a cotton ball. Place sachets or cotton balls around the house, particularly in areas that attract flies.

HOMEMADE FLY PAPER

Create your own homemade fly paper and hang near doorways or other areas where flies congregate.

¼ cup corn syrup
2 tablespoons sugar
2 tablespoons water
Paper grocery bag, butcher paper, or other sturdy paper
Yarn or twine

Mix corn syrup, sugar, and water together in a saucepan; heat gently until sugar dissolves. Cut paper into strips about 2 inches x 10 inches (or whatever size you desire). Using a hole punch or scissors, put a hole in the top of the strip and thread the yarn or twine through the hole; knot it in the hole, and make another loop to hang the strip. Drag paper through wet mixture, wetting both sides thoroughly. Hang the strips to dry over a cloth (to catch drips), and then hang wherever you want to catch flies.

FRUIT FLIES

Fruit flies are attracted to the apple cider vinegar smell, and the dish soap weighs them down, so they cannot escape the liquid.

APPLE CIDER VINEGAR TRAP

¼ cup apple cider vinegar
Dish soap
Water

Pour apple cider vinegar into a shallow bowl. Add a few drops of dish soap, and swish gently to distribute the soap. Add a small amount of water to help mix the liquids together. Place near the area where fruit flies congregate. Replace as necessary.

SPIDERS

Although spiders are our friends (they help keep our homes free of bugs!), there are areas where we'd rather not see them. Besides using one of the ideas that follow to repel spiders without chemicals, it is also important to clear away spider webs and egg sacs as soon as you see them, and to keep spider-prone areas free from the insects they eat. Spiders can taste what they touch, and they don't like the taste of several different plants. Mix up some Spider Spray to repel spiders in areas you need spider-free.

SPIDER SPRAY

10–20 drops essential oils (lavender, cinnamon, citronella, catnip and/or citrus oils such as lemon or grapefruit
2 cups water

Mix 10–20 drops total of one or more of the oils with the water, and pour in spray bottle. Spray as needed wherever spiders like to lurk: in corners, on the stairs, in the bathroom cabinets, and so on.

CATNIP-BASED SPIDER REPELLENTS*

Spiders don't like the smell of catnip, so find some creative ways to put catnip in the areas where spiders loiter.

Fresh catnip leaves
Water
Catnip essential oil (alternative)
Dried catnip leaves (alternative)

Make an herbal extract by grinding fresh catnip leaves and adding to water. Soak cotton balls in the mixture (or simply use catnip essential oil).

Or create a catnip sachet with dried leaves and some thin natural fabric. Put the cotton balls or sachets wherever you see spiders.

* Catnip is also a deterrent for mosquitoes, but it needs to be refreshed regularly.

MINIMIZING BUGS IN YOUR YARD

If the bugs aren't there, they aren't going to bite you. Eliminating or minimizing the annoying insects in your yard (only the annoying ones, not the others!) can go a long way toward enjoying your yard. Here are some simple steps to discourage vexing insects from invading your outdoor living space:

Eliminate Standing Water

Standing water is the perfect breeding ground for mosquitoes. Check your yard for sources of standing water, such as planter saucers, children's toys, clogged gutters, trash, or pet bowls. If you have a bird bath, change the water frequently. Ponds should have some type of pump to keep the water moving. Make certain any recycling bins have drainage holes.

Mosquitoes can lay eggs in as little as ¼ inch of standing water.

Attract Bats

Bats can catch between six hundred and one thousand insects per hour, putting a big dent in the flying insect population in your backyard. Although a small percentage of bats carry rabies, just as any other mammal can, bats pose little threat to humans who do not handle them. Consider purchasing or building a bat house for a little free bug-hunting labor.

Cut the Weeds

Adult mosquitoes, along with other insects, like to hang out among overgrown plants. Keep the weeds at bay to avoid giving bugs a handy place to hide.

Introduce More Insects

It might sound counterproductive, but introducing beneficial insects can help keep the population of biting insects in check. Ladybugs,[219] green lacewings,[220] nematodes, dragonflies, and praying mantises[221] are a few examples of beneficial insects. Release these insects directly into your yard, or plant flowers and native plants to encourage them to populate it on their own. Try planting some wintergreen. Its recognizable odor comes from its oil, methyl salicylate, and attracts a variety of beneficial insects.

> Repel insects naturally by planting beneficial plants. Plants such as citronella grass or lemon balm help keep mosquitoes away. Mint planted around your home helps keep ants from moving inside.

PLANTS THAT WARD OFF INSECTS

What's growing in your yard? You can affect the ecology of your yard with some careful planting. Many plants naturally repel insects or encourage beneficial predators of the insects you want to keep out of your outdoor living spaces.

BASIL (*OCIMUM BASCILICUM L.*): Flies can be a problem in warmer months, but some strategically placed basil plants will help keep the flies away, and a leaf or two of the plant will liven up your cooking. Beside its antifungal and antimicrobial properties, studies have shown that the essential oil in basil has strong insect-repelling abilities.[222]

BAY LEAVES (*LAURUS NOBILIS*): Known for their savory properties in cooking, bay leaves are also good at keeping away insects, such as cockroaches, silverfish, and moths. If you live in a climate where cockroaches are always trying to invade your home, try planting bay plants around your building to help curb invasions.

BEE BALM (*MONARDO SPP.*): Also known as horsemint, oswego tea, or bergamot,[223] bee balm has a scent that is described as a cross between mint and oregano. The plant grows wild throughout most of the eastern United States. Besides being a natural mosquito repellent, bee balm is also a natural fungicidal and bacterial retardant. When planted in a home garden, the plant attracts natural predators, such as birds, pollinating insects,[224] and predatory insects.

CATNIP (*NEPETA CATARIA*): Catnip is not just for driving your cats wild. The oil in catnip, nepetalactone, is ten times better at repelling mosquitoes than DEET.[225] Plant some catnip in your yard and around patios and walkways to help repel mosquitoes, flies, cockroaches, and more.[226, 227]

CHRYSANTHEMUMS (*CHRYSANTHEMUM SPP.*): These attractive plants, so popular in fall, repel roaches, fleas, ticks, bedbugs, lice, silverfish, ants, and more, as they contain pyrethrins, an important natural source of insecticide.[228] Pyrethrins attack the central nervous system of an insect and inhibit female mosquitoes from biting. In smaller amounts, they effectively repel insects.

CITRONELLA GRASS (*CYMBOPOGON NARDUS*): Citronella is the most common natural ingredient found in mosquito repellents — both natural and chemically based. Although many products that contain citronella are on the market, such as candles, torches, and scented plants, the living citronella plant has a much stronger smell, causing it to be a much more effective insect repellent.[229] Growing as large as six feet tall, citronella grass not only makes a natural insect repellent, it also makes an easy natural screen, giving privacy while needing very little care.

EUCALYPTUS (*EUCALYPTUS SPP.*): Eucalyptus trees require a significant amount of water and are particularly successful in areas with poor drainage, as they not only repel insects but also remove water that could otherwise attract insects.[230, 231] Lemon eucalyptus oil has been shown to work as effectively as DEET in repelling insects,[232] and it is effective for longer than many other naturally derived oils.

FLOSSFLOWER (*FLOSSFLOWER SPP.*): Also known as bluemink, flossflower is particularly offensive to mosquitoes. The flowers are easy to grow, and they thrive in both partial and full sunlight. The leaves of the flossflower plants can be crushed to increase the emitted odor, but take care not to transfer the oil in the leaves to your skin to avoid irritation.

GARLIC (*ALLIUM SATIVUM*): Known for its culinary attributes, garlic has long been employed for medicinal and other uses. Garlic has been shown to be an environmentally friendly deterrent to insects.[233]

GERANIUMS (*PELARGONIUM GRAVEOLENS*): The colorful blooms of geraniums will brighten the exterior of any home. Geraniums are sometimes called mosquito plants for their ability to repel insects. Geranium essential oil contains more than fifty organic compounds, including significant proportions of citronellol, nerol, and geraniol, known insect deterrents.[234, 235, 236]

LAVENDER (*LAVENDULA ANGUSTIFOLIA*): Lavender has been used as an insect repellent for many centuries, protecting cloth from infestations, repelling insects, and treating bites and stings.[237] Lavender plants are not only attractive but can also help deter insects from invading your yard. Break off some of the plants or dry the flowers to add to sachets in your home to help keep bugs out of your belongings.[238]

Lavender oil, in addition to helping prevent bites, can help in soothing insect bites and stings.

LEMON BALM (*MELISSA OFFICINALIS*): Lemon balm is in the mint family and has a subtle lemon scent. When the leaves are crushed, the essential oil in lemon balm, which contains the natural compounds citronellal, geraniol, linalyl acetate, and caryophyllene, is released and acts as a natural insect repellent.[239]

LEMONGRASS (*CYMBOPOGON SPP.*): Lemongrass, which has a fresh, lemony scent, is very effective at repelling mosquitoes.[240] Lemongrass can also be used inside your home.[241] The dried leaves make a lovely tea, and the oils are great for cleaning, leaving a light lemon scent behind. Lemongrass is also avoided by bees, so plant it in areas where you don't want bees, such as near your house.

Keep the bees in other areas, where they can busy themselves making honey.

LEMON THYME (*THYMUS SERPYLLUM*): Lemon thyme has about two-thirds the mosquito-repelling power of the high-powered and highly toxic DEET. Unlike many of the garden plants that help repel bugs, simply having lemon thyme in your yard alone will not keep the bugs away. The plant works best when the leaves are crushed and the plant's oil is used on the skin. Lemon thyme is a creeping plant, best placed as ground cover.

MARIGOLDS (*TAGETES SPP.*): Marigolds come in a variety of warm hues, from yellow to red. Although these annuals are attractive to look at, they emit an unpleasant scent that is repellent to many bugs, from mosquitoes to insects that attack plants. Because of this, marigolds are a perfect addition to your backyard vegetable garden. You can ward off the insects that could attack your vegetables while protecting yourself from becoming an insect feast.

MINT (*MENTHA SPP.*): Found in multiple varieties from spearmint to peppermint, mint has a light, refreshing scent. Useful in cooking, mint is also effective at repelling mosquitoes. Place the plants in your landscaping or rub the oils on your skin to help keep the mosquitoes at bay. Mint will also ward off other insects, such as ants, when placed around the perimeter of your home.[242]

PINEAPPLE WEED (*MATRICARIA DISCOIDEA*): Pineapple weed, a wild relative of chamomile, is an annual native to North America and Northeast Asia,[243] commonly found in lawns, by the side of the road, and in other areas where humans have disturbed the natural environment. The plant is aromatic when crushed and smells like pineapple, hence its name. The aromatic oils can help repel insects.

ROSEMARY (*ROSMARIUS OFFICINALIS*): Rosemary is a common herb grown by many for its uses in the kitchen. The oil in the rosemary plant is repulsive to many insects, including mosquitoes and spider mites.[244] The plant is suited to hot, dry climates with well-drained soil; it is more difficult to grow in colder climates. Rosemary is actually a shrub, so trimming is recommended to keep the plants from becoming too large.

VANILLA LEAF (*ACHLYS TRIPHYLLA*): Vanilla leaf is found growing in shady, moist areas. It's no coincidence that this plant grows in the same conditions that attract mosquitoes. The plant has long been used by indigenous cultures to ward off mosquitoes and flies; bundles of the dried plant are hung in homes.[245]

WORMWOOD, SAGEBRUSH, AND MUGWORT (*ARTEMISIA SPP.*): This genus of plant is an effective mosquito repellent when used as an aromatic smudge or when its oil is applied directly to the skin.[246] The plant is best suited to dry climates.

DIATOMACEOUS EARTH

Diatomaceous earth, or diatomite, is a naturally occurring sedimentary rock. It easily crumbles into a fine white powder, which is generally the form available for purchase. Look for food-grade diatomaceous earth online (Amazon.com carries it).

Due to its abrasive and physico-sorptiveproperties, diatomaceous earth is used as an insecticide.[247] In its fine-powdered form the substance absorbs the lipids on the waxy outer layer of an insect's exoskeleton, essentially causing the insect to dehydrate. The resulting water pressure deficiency then causes the insect to die.

Scattering diatomaceous earth throughout your yard can cut down on the number of insects invading your living space. Though medical-grade diatomite is sometimes used as a dewormer for humans and animals alike,[248, 249] care should be taken when applying it to avoid inhaling the fine particles.

NATURALLY HEALTHY YARDS

Sometimes referred to as greenscaping, synergistic outdoor spaces utilize biology for healthier yards. By considering the biological needs of plants, you can create a stronger and less toxic environment.

Save time by using plants that require less care. Save money by using fewer resources.

How can you change your yard to incorporate natural ways to make it healthy and avoid toxic chemicals near your home?

Protect the environment by using less water, generating less yard waste, and eliminating or greatly reducing the use of toxic chemicals.

HEALTHY SOIL

A teaspoonful of healthy soil contains approximately four billion organisms!

At the bottom of any yard is soil. Plants receive many nutrients from the soil. Well-nourished plants are healthier plants, which are able to better protect themselves from pests and disease.

Soil needs nitrogen, phosphorus, potassium, and lime to grow healthy plants. Feeding your soil with compost is a natural way to add nutrients to your yard or garden space. Compost can be tilled into new garden or flower beds or areas where new grass is being planted. It helps sandy soil hold nutrients and water, loosens compacted clay soil, and feeds the beneficial microorganisms living below your feet.

Take it slow with fertilizer. One of the easiest ways to damage your yard is to add too much or the wrong kind of fertilizer. If your lawn does need some added nutrients, look for labels such as "natural organic" or "slow-release." These products do a better job of feeding your plants slowly and evenly over time, strengthening root systems for healthier plants.

MULCH

Mulching — adding a layer of organic material, such as leaves, aged wood chips, or grass clippings — can help stabilize soil temperature, prevent weeds, feed the soil, and conserve water. If you use organic material, it will decompose over time, recycling itself into your soil! Add mulch to flower beds and vegetable gardens or around trees and shrubs, but avoid adding more than three inches of mulch to an area, and leave space in between mulch and plant stems or tree trunks. Can you mulch your lawn? Try grass-cycling! Grass clippings left in the yard will quickly decompose, releasing valuable nutrients back into the soil without contributing to thatch, an over-abundance of dead and living roots of plants that prevent aeration and proper drainage.

 Skip bagging your grass clippings. Instead, leave them on the ground to save time and reduce your lawn's need for fertilizer.

DIVERSE LAWNS

Lawns with only one type of grass may look healthy, but they really aren't. Monoculture lawns are more susceptible to pests, more vulnerable to environmental conditions and changes,[251] and require more resources, whether water, fertilizer, or pesticides and herbicides. The key to a healthy lawn is biodiversity.[252] Biodiversity allows species, individually and as a whole, to benefit from one another. The healthier lawns are the ones in which multiple species exist.

Decrease weeds, reduce the need to mow, and add nutrients to your yard by diversifying your lawn. Adding clover or other beneficial grasses will result in a healthier yard, which is easier to care for.

GRASS SEED

When you purchase grass seed, you may notice that the labels on most bags state that the contents are either a mixture or a blend. These words do not mean the same thing. Mixtures contain more than one type of grass seed species. The label should list the percentages of different species by weight. Blends contain only one type of grass species. These are hybrids that have been bred in order to incorporate the beneficial qualities of two or more species of grass. Overall, a mixture will do more to increase the genetic diversity of your lawn, making it healthier and increasing its resiliency.

CLOVER LAWNS

Many people consider clover a weed, but that wasn't always the case. Until the rise of broadleaf herbicide in the 1950s, a healthy patch of white clover was actually considered a standard of excellence in lawn care.

Clover lawns have some advantages over traditional grass lawns:

- **CLOVER STAYS GREEN.** In most regions of the United States, clover will stay green with little or no watering, being relatively drought-tolerant. It turns green early in the spring and remains so until the first frost.

- **YOU DON'T HAVE TO MOW CLOVER.** White clover grows only 2–8 inches tall. Little mowing is needed to keep a clover lawn looking neat. Sometimes a midsummer mow to deadhead old blooms is all you need!

- **NO FERTILIZER IS NEEDED.** Clover is a nitrogen-fixing legume, adding nitrogen back into the soil, which benefits surrounding plants. You will never need to add fertilizer to clover, and any grass planted with the clover will be healthier and green. Clover does well in most types of soil conditions.

- **CLOVER KEEPS OUT WEEDS.** Clover is a very persistent plant. If you are trying to avoid other weeds in your yard, you are in luck.

Clover easily out-competes most other weeds, eliminating the need for toxic herbicides or weeding by hand.

- **CLOVER IS DOG-SAFE.** If you have had problems with brown patches in your yard from dog urine, try planting clover in those areas. Clover is not killed by the urine; it stays green and lush.

- **CLOVER IS INEXPENSIVE.** Compared with the price of grass seed, clover is extremely inexpensive. If clover is trying to invade your yard, you are getting it for free. If not, you can find bags of clover seed for a few dollars.

There are a few disadvantages to a clover lawn:

- **CLOVER DOES STAIN CLOTHING MORE EASILY THAN GRASS DOES.** Plant clover in areas where children are less likely to play rough and tumble games.

- **CLOVER IS LESS DURABLE THAN SOME SPECIES OF GRASS.** On it's own, clover is not durable for high-traffic areas, such as playing fields, although it does quite well when mixed with grass.

- **CLOVER IS A PERENNIAL AND IS SHORT-LIVED.** For clover-only yards, you may have to reseed every 2–3 years. However, in a mixed yard, clover will reseed, reestablishing its presence on its own.

ROUND UP THE WEED KILLERS

There are a multitude of herbicides used to kill unwanted plants, and they carry a multitude of health and environmental risks, despite claims from their manufacturers.[253, 254] The first herbicide was 2,4-dichlorophenoxyacetic acid (2,4-D), although it began as part of the U.S. military's chemical warfare program (it was a component of Agent Orange).[255] Scientists are divided concerning the carcinogenic attributes of the chemical, and other studies have shown clear links between exposure and increased illnesses, such as non-Hodgkin's lymphoma[256] and amyotrophic lateral sclerosis.[257, 258] One study found occupational exposure to 2,4-D to be associated with male reproductive problems.[259]

The use of the triazine family of herbicides, which includes atrazine, has been controversial due to its widespread contamination of drinking water and possible associations with birth defects and reproductive issues.[260] The use of atrazine was banned in the European Union in 2004 due to persistent groundwater contamination,[261] but it remains one of the most widely used herbicides worldwide, with 76 million pounds of the chemical applied annually.[262] Glyphosate, the key ingredient in many commercial weed killers today (Monsanto's Roundup, for example), is the most widely used herbicide in the world.[263] Glyphosate has been shown to cause damage to and/or disrupt the endocrine system, the liver, and other bodily systems.[264]

GETTING RID OF WEEDS NATURALLY

How can you fight weeds without resorting to toxic herbicides?

Eliminate weeds naturally by pulling them or by using boiling water, rock salt, vinegar, baking soda, or other natural methods.

PULL WEEDS: Try pulling weeds by hand. This is more practical if weeds are sparse or localized. Some weeds are harder to pull than others. Try pulling after a good rain or after watering the area to make the ground softer. It may help to insert a screwdriver or blade against the root to help pry it loose before pulling. Grip firmly at the base and pull so that the root comes free.

BOIL WEEDS: Scalding water will have most weeds shriveling in a couple of days. Try repurposing boiling water from your cooking or boil some fresh water.

SMOTHER WEEDS: If you are trying to clear an area, such as for a garden plot, cover the area with newspaper or cardboard first. The lack of sunlight will help kill the weeds and the paper will naturally degrade over time.

USE SALT: Use rock salt on areas, such as gravel drives, where you don't want weeds. Salt also makes a good weed barrier along lawn edges, where it is difficult to mow. Be careful near walkways, as salt will erode concrete surfaces.

BLOCK WEEDS: Lawn edgings and retaining walls can act as a weed-break. If you are fighting against weeds coming in from the woods, try installing a fence. If the seeds can't get there, the weeds won't grow.

OUT-COMPETE THE WEEDS: All living species need resources. By planting clover, flowers, and gardens, you can take away their resources, greatly reducing the number of weeds.

USE ACID: Pouring vinegar onto plants is an easy way to kill them. This is perfect for the edges of a lawn or the spaces between walkways, where you can easily control your vinegar application. Try straight white distilled vinegar or reach for the leftover juice from your finished pickle jar.

APPLY CORNMEAL: Sprinkling cornmeal on the ground will prevent seeds from germinating. Be sure to wait until any plants you do want are growing. As a side benefit, cornmeal attracts earthworms, which loosen and fertilize soil.

EMPLOY BAKING SODA: Baking soda changes the pH of the soil, making it difficult for plants to grow. Try this in areas where you don't want any plants growing.

CUT BACK ON GRASS

Many communities have started initiatives to get back to their roots by replacing lawn with food.

Tired of mowing grass? Consider cutting back on how much grass you have. Rather than fighting to have that perfect lawn, consider adding visual appeal by replacing a portion of your grass. Rain gardens, which utilize native plants, are a fantastic way to add appeal, minimize standing water, and strengthen the ecosystem in your yard. Areas that prove difficult for grasses to grow may be the perfect places for ground cover. Other areas may benefit from an edible garden.

PLANT FOR YOUR SPACE

You can plant whatever you want in your yard, but if the conditions aren't right, you will find yourself fighting a losing battle. For plants that grow well in your area, look at native species, which have evolved to live in a particular area. Because they are well-adapted to their environment, they require less maintenance and added nutrients. They give back to the local ecosystem, providing food and habitats appropriate for local wildlife, which then help in pollination and by removing harmful insects. Leaving wild buffer zones of native plants along wild areas (including streams or woods) and along fence lines will help keep out invasive plants, such as poison ivy, typically found where humans have disturbed the natural environment.

> Using native species in your landscape will result in easy-to-grow plants and a healthier yard, which requires fewer added nutrients and less water.

MAKE EVERY DROP COUNT

Too much water can be just as much of a problem as too little water. Most plants do better if the soil is allowed to dry partially between waterings. Although some plants require watering at the first sign of wilting, mature trees and shrubs generally do just fine without watering once they have established their root systems, with the exception of extremely dry years.

Lower your water bills while providing more water to your plants:

- Use native plants well adapted to the area's resources.

- Reduce evaporation through the use of compost and mulch.

- Use soaker hoses or drip irrigation in plant beds, as they are more efficient than sprinklers.

- Watch where your water is going. Watering the pavement wastes resources.

- Stop watering if puddles begin to form. Wait for the water to be absorbed into the soil before continuing watering. This will prevent runoff, which is bad for your yard.

- Water in the early morning. Watering at midday results in a high degree of water evaporation. Watering in the evening can cause mold growth and plant diseases.

- If you are in a dry spell, allow the lawn to go dormant. Minimal watering will still allow brown areas to bounce back in the fall.

- Catch rain water in rain barrels and through properly directed downspouts.

COMPOSTING

Compost is simply organic matter that has decomposed. It makes an excellent fertilizer, adding various nutrients to soil to give your garden, flowers, or yard a healthy advantage. There are many ways to successfully compost, and they all use items you are currently throwing away, including yard waste and food scraps. Natural composting, or biological decomposition, has existed throughout Earth's history and continues to be the normal process by which organic materials break down and are reused by the Earth. With a few easy steps, you can begin composting and turn trash into a veritable wealth of nutrients for your outdoor spaces.

Reduce or eliminate the need for commercial fertilizer in your yard or garden by starting an easy compost pile.

THE BENEFITS OF COMPOSTING

Wondering why you should compost? Check out the benefits of composting your waste:

- **REDUCE SOIL'S NEED FOR WATER, FERTILIZERS, AND PESTICIDES.** When you naturally make the soil healthier and more nutrient rich, your plants flourish without chemical products and without you spending a penny.

- **KEEP ORGANIC MATTER OUT OF LANDFILLS.** Not only are you limiting the amount of items in landfills, you are allowing the organic matter to break down. Due to the way landfills are constructed, they rarely have enough oxygen to allow microorganisms to break down organic matter. It just sits there, taking up space and not helping the Earth.

- **IMPROVE THE SOIL.** Increase the activity of soil microbes, improve soil chemistry, and improve insect and disease resistance in your plants and trees.

- **DECREASE YOUR USE OF RESOURCES.** It may seem that it doesn't cost anything to wash food down the garbage disposal or put it in the trash. But water that needs to be treated and trash that has to be transported costs money and environmental resources. The costs to the environment may be hidden but are no less real.

MICROORGANISMS AT WORK

If composting is so easy, who is doing the work? Microorganisms, too tiny to see with the naked eye, are hard at work in your compost pile or bin. When you put food and yard scraps into your compost, you source an entire food web. Bacteria and fungi break down the organic matter. Protozoa, single-celled organisms, nematodes, and mites feed on the bacteria and fungi. Invertebrates, along with predatory nematodes and mites, keep the protozoa, mites, and nematodes in check. Together, these organisms form a balanced ecosystem that makes composting a natural and efficient process. The organisms break down the organic waste into its simplest parts, leaving behind nutrient-rich humus, a mature form of compost that won't break down any further.

WHAT CAN YOU COMPOST?

You might be surprised by the sheer amount of items that can be composted in your own backyard.

- **FRUITS AND VEGETABLES.** From leftovers to peels, skins, and leaves, toss your fruit and vegetable scraps into the compost.

- **EGG SHELLS.** Crush up egg shells to add nutrients to your decomposing waste. Because egg shells are harder than other items, crushing them allows the bacteria to more effectively break them down.

- **COFFEE GROUNDS AND TEA BAGS.** If you drink coffee or tea, you're in luck. Those coffee grounds and tea leaves, along with the paper filters and bags, can be tossed into your compost.

- **YARD WASTE.** From grass clippings to leaves, yard waste is easily composted and adds valuable nitrogen to your compost bin. However, too much can add an excess of nitrogen to the compost, making the process less effective. If you have a large yard or farm, consider adding a different compost pile for your yard waste.

- **PAPER RECYCLING.** Recycling paper is great, but sometimes paper can't be recycled, or perhaps you have shredded important papers and don't trust those pieces to someone else. Try adding paper to your compost bin.

- **SAWDUST AND STICKS.** Whether you have small sticks in your yard or are an avid woodworker, wood makes an excellent source of carbon. Small pieces or sawdust are more easily broken down in a compost bin.

WHAT SHOULDN'T YOU COMPOST?

You can compost many things, but there are some organic items that shouldn't be added to your compost.

- **MEAT AND DAIRY.** Decomposing fruits and vegetables have a slight smell, but it is nothing compared with the stench of rotting meat and dairy products. Not only do these items stink, but they can also attract pests to your compost area and your outdoor, or even indoor, living space.

- **HUMAN AND PET WASTE.** It is true that decomposed manure makes an excellent fertilizer, but feces has some additional considerations for its decomposition. Special care needs to be taken when decomposing animal waste, and there are definitely ways to do it safely. If you plan to compost human or pet waste, read up on how to properly do so to prevent the spread of disease and parasites.

- **DISEASED GARDEN PLANTS.** If you put diseased garden plants into the compost, they can infect the rest of your compost.

- **INVASIVE WEEDS.** Although composting breaks down organic material, some seeds and spores can survive the compost process. Adding what you think is nutrient-rich compost humus to your garden or yard, only to find out that it is invasive weed seeds, would be a disappointing discovery.

- **CHARCOAL ASHES.** Do you have charcoal ashes from your grill or fire pit? They have carbon, but you will want to avoid adding these to your compost bin. The ashes are toxic to soil microorganisms and will kill the microecosystem living in your compost.

- **PESTICIDES.** These products were developed to kill pests and can also harm the microorganisms living in your compost. Not only should you not spray pesticides into your compost, you should avoid adding plant material that has been treated with pesticides. The great thing about using compost in your yard is that you are cutting down on the need for chemical pesticides and fertilizers.

CONTAINING YOUR COMPOST

Once you have made the decision to compost, you have to consider where you will put your compost. Most backyard composters use one of two methods.

Those looking for the least amount of work and expense or who are just getting started may choose the pile method. Organic material is placed in a pile and left to decompose. You can choose to turn the pile to aerate it or leave it to its own devices. As the material begins to decompose, the pile will get smaller.

The other method is to put the compost in some type of container. You will find compost containers in all price ranges and styles, from an expensive commercial compost bin with handles to turn the barrels to a homemade recycled container with holes drilled in the bottom to prevent excess water from sitting and becoming stagnant.

There are many options when it comes to composting, and you can begin composting immediately. Consider collecting your kitchen scraps in a bowl and taking them out at the end of the day. Before long, you will have rich humus to add to your flowers, garden, and yard.

VERMICOMPOSTING

Vermicomposting, or worm composting, can reduce composting time by as much as 50 percent. With this method, organic matter is placed in a bin containing red worms or earthworms. The worms consume the material, breaking it down into a nutrient-rich compost called worm castings.

Vermicomposting has very few basic requirements:

- Worms
- Worm bedding, such as shredded newspaper or cardboard
- A properly vented bin with drainage
- Food scraps for the worms to consume

Worm bins can be purchased premade. However, many easy-to-follow tutorials exist to explain how to make your own vermicomposter. Worm bins are a great learning opportunity for children, showing children how food scraps are broken down by the worms as food for plants.

One pound of worms can eat as much as one-half pound of organic material per day.

FOR PARENTS

For many consumers, the motivation to move away from commercial cleaning products is born at the same time as their first baby. In this section, you will be introduced to a few good reasons for parents to move away from toxins, along with some recipes particularly relevant to parents.

KEEP OUT OF REACH OF CHILDREN

Many toxic cleaning products smell fruity and sweet and are wrapped in colorful packaging. To make matters worse, many cleaners do not have childproof caps and are easily accessible to little fingers. These dangerous chemicals can look just like a juice or sports drink, and kids who cannot read (or who don't take the time to look) might take a gulp. Many of these same cleaners also carry a label warning parents to "keep out of reach of children," usually because of problems that can arise if children ingest them.

Risks of chemical exposure for children are not limited to ingestion, though. Children also are exposed to toxic chemicals through inhalation and skin absorption. One study showed early exposure to the chemicals found in common household cleaning products was linked to a 41 percent increase in a child's chances of developing asthma by age seven.[265]

The average American uses approximately 40 pounds of toxic household cleaning products each year, products our children are in constant contact with when the products are on the floor, on other surfaces, on toys, and in the air.[266] That teething ring you cleaned off with a Clorox wipe? It just went into your little one's mouth. The carpet cleaner you used to remove the stains left by your toddler? It gets into his skin when he sits on it and plays with toys that spend time on the floor. The coffee table you cleaned with Pledge? Your baby pulls up on it and then innocently gums it.

Cleaning products consistently rank as one of the top five causes of pediatric poisoning.

Children have a greater exposure to chemicals than adults when comparing mass size.[267, 268] Children are also more vulnerable to chemical exposure because their organs and immune systems are not yet fully developed.[269] Chronic, low-level exposures to some of the chemicals found in common cleaning supplies pose a significant health risk to children, including an increased incidence of asthma, allergies, some types of cancer, learning and behavioral disorders, endocrine disruption, chemical sensitivities, and kidney or liver damage.

RECIPES FOR PARENTS

All the recipes in this book will be helpful to families with children, but there are a few recipes that may be more useful once you're cleaning with little ones afoot.

Care should be taken to properly store all items, including those made from natural ingredients, if you have children or pets. Just because a product isn't labeled "Keep out of reach of children" doesn't mean it's safe for your kids. And as with our other recipes, you may want to test any of the recipes that follow on an inconspicuous area first.

STICKER REMOVER

White distilled vinegar

Are you tired of working hard to remove stickers or labels from hard surfaces? Try soaking the sticker in white distilled vinegar, allowing it to sit, and then washing with a wet, soft cloth. Repeat to remove any sticky residue.

PERMANENT MARKER REMOVER

Vodka

Hard surfaces: Using vodka and a soft cloth, gently rub the area to remove permanent marker from hard surfaces.

Carpet: Follow the preceding directions, but blot with a towel instead of rubbing.

CRAYON REMOVER

Lavender or tea tree essential oil
Dish soap
Baking soda

Hard surfaces: Using lavender or tea tree oil and a soft cloth, gently rub the area to remove crayon from hard surfaces. Follow by gently rubbing with mild dish soap and a soft cloth. You may want to test an inconspicuous area first.

Carpet: Remove excess crayon with a knife. Place a clean paper bag over the crayon, and run a warm iron over the bag. As the crayon transfers to the bag, use a clean part of the bag to remove more crayon. Follow by sprinkling the crayon stain with baking soda, and gently rub with a sponge or soft cloth. If needed, finish by blotting the stain with tea tree oil.

GUM REMOVER

Ice
Orange or lemon essential oil

Carpet or clothing: Put ice in a plastic bag, and then apply to gum. Remove as much of the cold gum as possible with a butter knife or your fingernails. Then, using orange or lemon essential oil on a cotton ball, gently scrub at the remaining gum. The International Chewing Gum Association explains that natural solvents in citrus oil can help remove gum.[270]

CHOCOLATE REMOVER

1 tablespoon borax
1 cup warm water
Dish soap

If the chocolate is dry, vacuum up all loose bits. Dissolve borax in warm water. Using a soft cloth, gently rub the stain with the borax mixture until it is clean. If necessary, follow by gently rubbing with mild dish soap and a soft cloth.

DIRT AND MUD REMOVER

Dish soap
1:1 solution hydrogen peroxide and water

Allow wet mud to dry and then scrape and vacuum up as much dried dirt as possible. Using dish soap, water, and a soft cloth, gently scrub at the stain. If necessary, follow with a 1:1 solution of hydrogen peroxide and water.

FECES, URINE, AND VOMIT REMOVER

Hydrogen peroxide

Remove as much of the solid material and excess liquid as possible. Pour hydrogen peroxide onto the stain and blot with a soft cloth. Continue blotting and gently rubbing until stain is gone.

If at all possible, allow sun to shine on carpet — sun is an excellent organic stain remover.

HEALTHIER ALTERNATIVES FOR COMMON CHILDREN'S PRODUCTS

CLOTH DIAPERS

Disposable diapers contain chemicals that affect our babies' health. Fragrances and dyes in many diapers are associated with respiratory problems, skin irritations, and increased asthma rates.[271] Disposable diapers also contain three chemicals that have sparked criticism: dioxin, sodium polyacrylate, and tributyltin.

Dioxin is a concern for both disposable and cloth diapers manufactured where chorine-bleaching processes are used. Sodium polyacrylate, the chemical that makes disposable diapers absorbent, can hold as much as three hundred times its own weight in liquid. In the 1980s, sodium polyacrylate was removed from tampons after it was linked to toxic shock syndrome. Sodium polyacrylate has been linked to genital bleeding, fever, vomiting, and staph infections in babies, and it can be fatal if ingested.[272, 273]

Tributyltin (TBT) is a chemical that was at one time used in paint on boats, where it was used to prevent barnacles from attaching to the hulls. An endocrine disruptor, its use was discontinued when research demonstrated its serious effects on marine life; in one aquatic species, it caused the females to develop male genitalia, thereby becoming unable to release their eggs. It is surprising that a chemical known to be absorbed through the skin and to affect sex hormones would be found in many diapers, products that sit right up against our babies' sexual organs.[274]

With the exception of dioxin, all of the chemicals mentioned here are absent from cloth diapers. Aside from the health risks, there are environmental consequences as well. Every child in disposable diapers will add an average of almost 3,800 diapers to our landfills in roughly 2.5 years.[275] That translates to more than three million tons of diaper waste dumped in landfills each year in the United States alone.[276]

There are many options available for cloth diapering. For families on a tight budget, check into unbleached cotton prefolds or flat diapers coupled with waterproof or wool covers. You can find other affordable options by buying "seconds," which are new diapers with slight defects offered at deeply discounted prices from the manufacturer. The defects will be aesthetic, rather than functional. Alternatively, check into the cloth diapering communities online for gently used cloth diapers and accessories.

Cloth diapers may seem like a bigger investment at the beginning, but the money needed to start cloth diapering is small compared with the thousands of disposables you would buy in the long run. Additionally, cloth diapers can be reused with subsequent children or resold to recoup even more of the investment.

BABY WIPES

Baby wipes are known to contain several types of toxic chemicals. In addition to dioxin, commercial wipes contain phthalates, which are linked to reproductive problems. Parabens — which can interfere with the endocrine system, can cause skin irritations and cancer, and are linked to learning disorders, reproductive and developmental disorders, and immune system dysfunction — are another ingredient. Lastly, wipes can contain propylene glycol, an industrial antifreeze and an ingredient in brake fluid, which can irritate skin and mucous membranes.[277, 278, 279, 280]

But you don't need a plastic container with chemically moistened wipes to clean dirty fingers or bottoms. All you need is a piece of cloth and something to wet it with.

What to Use for Cloth Wipes

You have a few options for cloth wipes. Start off by cutting up old, soft T-shirts or invest in some flannel.

REPURPOSED T-SHIRTS: Cut T-shirts up into approximately 6-inch x 8-inch rectangles.

FLANNEL: Using old 100 percent cotton flannel baby blankets or a length of flannel from the fabric store, cut 6-inch x 8-inch rectangles. Now you have a decision to make: one ply or two ply? One ply is sufficient, but you might want to cut one-ply wipes with pinking shears to help avoid fraying. Two ply provides increased durability. The simplest method is to place two pieces of flannel together and serge around the outside. If you don't own a serger, put two pieces of flannel together with the right side facing in, and stitch around the outside edges, leaving a gap on one side. Turn the wipe inside out, and top-stitch around the edges, closing that gap.

Baby Wipes Solution

Following are a few recipes for wipe solutions for family cloth and baby wipes. In each solution, you'll find water, essential oils (such as tea tree oil for its antiseptic properties and/or another oil for fragrance or healing properties), a fruit- or vegetable-based oil (for softness and comfort), and/or soap for cleaning. You can also use a water–and–witch hazel mixture. Witch hazel has skin-healing properties but can sometimes sting broken or inflamed skin. You can play with the amounts and types of ingredients until you find a recipe that is best for your baby's bum.

Store extra wipe solution in a canning jar or other container, and keep the container in a cool, dark cabinet. Fill a spray bottle or a repurposed soap dispenser as needed, and spritz or pour onto wipes at each use. Pre-wetting your wipes is tricky, because wet wipes left too long can mildew. If you do prefer to pre-wet your wipes, do a handful at a time, and launder them after a few days.

Remember, though, you don't have to make a fancy wipe solution. Water is ordinarily just as effective.

SIMPLE AND SOOTHING WIPE SOLUTION

2 cups water
1–2 tablespoons olive or jojoba oil
2 drops tea tree essential oil
5 drops lavender essential oil or other essential oil

Mix all ingredients together and shake well before use.

 Replace commercial baby wipes with cloth wipes and a homemade wipe solution.

SOFT AS A BABY'S BUM WIPE SOLUTION

2 cups of a 1:1 ratio of warm water and witch hazel
1–2 tablespoons apricot or avocado oil
1–2 tablespoons aloe vera gel
5 drops essential oil of your choice (try geranium, lavender, or Roman chamomile)

Mix all ingredients together and shake well before use.

GENTLE CLEANING WIPE SOLUTION

2 cups water
5–7 drops castile soap
5 drops lavender essential oil or other essential oil

Mix all ingredients together and shake well before use.

WASHING BABY WIPES: Laundering your baby wipes is as simple as putting them in the wash. Throw them in with your regular loads of laundry. Although washing on hot is preferable, your cloth wipes can get clean along with other clothes. Consider adding 10–15 drops of tea tree oil to your wash cycle. Its antibacterial and antifungal properties will help get your wipes super clean.

If your baby is sick and the wipes are more soiled than usual, throw them into a prewash cycle with a squirt of detergent and some tea tree oil, then add your remaining laundry for the regular wash. If you want to keep the wipes separate in the wash load (so that they do not get mixed in with a roommate's laundry, perhaps), wash and dry them inside a mesh laundry bag.

If the wipes become stained and it bothers you, dry them outside in the sun.

PLASTIC DISHES

Although plastic dishes are durable, their multitude of potential toxins makes conscientious parents rule them out. Stick with traditional glass or ceramic dishes for everyday use. Tempered glass dishes can take a beating without breaking. Non-breakable options include stainless steel, natural lacquered wood, and bamboo. Some glass plate companies now have silicone sleeves that fit on their plates, making the plates less likely to slide and fall.

Glass is an excellent option for drinks, and thicker glasses can be found in smaller sizes for young children. The most toddler-friendly glasses are actually ones that are sized proportionally and are weighted properly so they don't accidentally tip over.

Stainless steel utensils can grow with your child without potential health risks. They won't break and can be easily washed free of germs and bacteria.

Plastic Straws

Straws seem relatively inconsequential. They are fun to use whether you are drinking through them or blowing bubbles in your beverage. But although they were once a novelty for children, our increased consumption of fast food has resulted in an exponential growth of plastic straw usage. Alternatives to plastic straws include reusable stainless steel and glass versions. You will be happy to know that reusable straws also have their own handy cleaning brushes to keep them pristine, and most glass straw companies include lifetime warranties with their straws. If your straw breaks, they will replace it. The old glass straws can be recycled into new glass.

Each year, enough plastic straws are produced to fill more than 46,000 full-size school buses.

A DISPOSABLE SOCIETY

Our society has become a disposable one; few products currently on the market are meant to last. As far as marketing goes, planned obsolescence, a company's decision to make products so that they will become out of date or useless in a certain period of time, is a great business strategy. Products that can be reused or are long-lasting mean fewer purchases and hence less revenue for manufacturers. Disposable products must constantly be replaced.

The cost of living in a throw-away society hits more than just your wallet. For every item that is manufactured, resources went into making it, including many that are not easily renewable. Those items then require more resources for packaging and transportation before making it to your door. For the most part, the packaging is thrown away, and when the item reaches the end of its short life, it, too, is tossed by the wayside. The chemicals used in every step of the process affect the world around us.

PAPER PRODUCTION AND IMPACT

The population of the United States represents only 5 percent of the world's population but uses one-third of the world's production of paper.

Paper is both biodegradable and a renewable resource. Yet the manufacture of paper has a considerable environmental impact: Paper's manufacture "is the third largest user of fossil fuels worldwide."[281]

The production of paper takes a considerable amount of chemicals, water, and electricity. There is also the added environmental impact of gaining access to raw materials from the forest (or recycling plant) and transporting them to the manufacturer to the store to the consumer. If we switched to reusable products, these transportation costs would be virtually eradicated.

Chemicals are used throughout the entire production route, from forestry to washing the pulp used to make paper. As a result, the paper industry emits about 365 million pounds of chemicals into our environment annually, affecting our water supply and everything it touches.[282]

Two of the most harmful chemicals released into our environment due to paper production are dioxin and methane. Dioxin is a by-product of the production process, and methane is formed when paper breaks down in either landfills or in compost.

The United States paper industry dumps about 1 million pounds of toxic chemicals into our environment every day, the equivalent in weight of 365 Boeing 747-8s in chemicals annually.

When mills use chlorine or chlorine compounds in their bleaching processes, the chlorine breaks down and creates other toxins, the most potent of which is dioxin.[283] Dioxin "is one of the most toxic man-made substances."[284] The primary substance in Agent Orange and a widely studied toxin, dioxin is linked to many health issues in humans, including birth defects, endometriosis, "disruption of normal hormone signaling pathways, reproductive and developmental defects, immune-toxicity, liver damage, wasting syndrome, and cancer."[285]

Once dioxin is released into the environment, it is persistent: It does not easily break down.[286] Our health is affected not only by dioxin in the environment, but also by dioxin present in paper products themselves.[287]

Methane is one of the three main greenhouse gases. As the amount of greenhouse gases increases in the atmosphere, global temperatures increase.[288] When paper rots or is composted, it emits methane gas. In the past two hundred years or so — since the Industrial Revolution — methane concentrations in the atmosphere have more than doubled.[289]

PLASTIC

Although plastic has been around for quite some time, we continue to learn more about risks associated with chemicals in various plastics. Understanding how and when to safely use plastics can cut down on their negative impacts so that we can enjoy their benefits.

Bisphenol A (BPA), an endocrine disruptor used in the production of plastics, constitutes one of the highest volumes of chemicals produced worldwide, with more than 3.6 million tons manufactured annually.[290, 291, 292, 293, 294] It is found in drinking containers and in the lining of most food and beverage cans, plastic food containers and cutlery, toys, dental products, and even paper products.[295]

In 2006 the United States sponsored an assessment, conducted by thirty-eight experts, of BPA in scientific literature.[296] Their consensus was that "BPA at concentrations found in the human body is associated with organizational changes in the prostate, breast, testes mammary glands, body size, brain structure and chemistry, and behavior of laboratory animals."[297] Despite these findings, the FDA denies that there are any safety issues related to BPA use and food. There are currently no BPA labeling requirements for plastics: buyer beware!

BPA is not the only chemical in plastic that can be harmful to our health and the environment. PET, HDPE, PVC, melamine, and other chemicals also pose risks. Polyethylene terephthalate (PET or PETE, recycling code 1) is commonly used to make drink bottles and plastic food bags; it leaches phthalates and antimony compounds.[298] High-density polyethylene (HDPE, recycling code 2) is made from petroleum and is commonly used to make plastic bottles, corrosion-resistant piping, and synthetic lumber. Many forms of HDPE contain BPA.

Plastic drink bottles account for 30 percent of the world's PET production. PVC is the third most widely produced plastic worldwide.

Polyvinyl chloride (PVC) can be rigid (for example, in construction materials made of PVC) or flexible (such as in shower curtains and plastic food wrap made of PVC). Most PVC products contain phthalate plasticizers, which are known endocrine disruptors. Melamine, a durable material commonly used for dinnerware for young children, represents several health concerns. When heat is applied, including from heated foods, melamine releases toxins into the food. Chronic exposure can result in cancer or reproductive damage.

Recently, with consumers shunning BPA, many manufacturers have turned to bisphenol S (BPS). But like BPA, BPS is known to have estrogenic activity. In addition, it is a persistent toxin and environmental pollutant.[299, 300, 301]

As consumers, we shouldn't be asked to gamble with our health and the environment. We need to demand that our government mandate more rigorous research in the development of products and that the government show more regard for consumers than for the profits of companies and trade organizations.

PLASTIC IN FOOD

The use of plastic with food represents a great health risk to individuals. Chemicals leach into food and drinks, which are then consumed.

To eliminate the risks of chemicals leaching into your food, choose alternatives to plastic, such as glass containers or reusable fabric bags.

Tips to Avoid Plastic and Minimize Its Risks

- Look for glass bottles or plastic bottles made from polyethylene or polypropylene.

- If you are giving your baby breastmilk or a breastmilk alternative in bottles, opt for silicone nipples instead of latex, which can leach carcinogenic nitrosamines.[302]

- Look for cups made of polyethylene or polypropylene plastics, or preferably thick glass or stainless steel.

- If avoiding polycarbonate (#7 plastic) completely is not an option, avoid heating food or drink in it, as this increases the leaching process.

- Discard the plastic if any signs of wear occur, such as scratching, cracking, or clouding.

- Oily and fatty foods also leach chemicals from plastic more quickly.

- Avoid plastic drinking bottles. If you do pick up a bottle of water made with #1 or #2 plastic, remember that these are made for one-time use. Recycle them rather than reusing them.

- Avoid drinking from water bottles that have been left out in the heat, even if you have not refilled them. Consider alternatives, such as glass or stainless steel water bottles.

- Exercise caution when using plastic wrap, especially when it is heated in the microwave. Use alternatives, such as waxed paper, or at least try to make certain it does not come in contact with your food.

- Avoid packaging, which is a major source of unregulated plastic in our society.

- Avoid using plastic dishes or utensils. Many alternatives exist, including glass, ceramic, wood, stainless steel, and lacquerware. If you don't want your child to use a breakable dish, consider stainless steel or wood that has been coated with nontoxic lacquer.

REDUCE AND REUSE

"Reduce, reuse, recycle" is the mantra of conscientious consumers, but we tend to overemphasize the recycling portion and de-emphasize the first two steps. Reducing consumption and reusing items are critical if we are to reduce our impact and make our homes, and our environment, a healthier place. It just so happens that by switching to reusable products, we automatically reduce our consumption.

TIPS TO REDUCE PACKAGING

Product packaging in the United States results in an incredible seventy-six million tons of waste destined for the landfill each year.[303] What's more, paper and plastic packaging contain many of the very toxins we are attempting to avoid,[304] without the regulation that applies to many of the end products contained within. Your toxin-free product may be contaminated by its packaging. We can make a large impact by reducing the packaging that comes with our purchases.

Containers and packaging represent 32 percent of all municipal solid waste in the United States.

BUY LESS: If you aren't buying as much stuff, you won't be using as much packaging. Consider purchases carefully, and reduce the number of unnecessary purchases. Save money and the environment by buying less.

BUY IN BULK: The packaging for most toilet paper, paper napkins, and paper towels isn't elaborate, but when you consider that the average American uses almost 21,000 sheets of toilet paper[305] and 2,200 napkins each year,[306] that adds up to a lot of multipacks, which are all conveniently wrapped in plastic. Opting to buy in bulk reduces the amount of packing used. Buying larger quantities may also help you save money (bulk items are often offered at a discount) and time (you will not have to shop as often).

PURCHASE FROM ECOLOGICALLY RESPONSIBLE COMPANIES: Choose to buy from companies who are making an effort to limit excess packaging. Look for companies who eliminate or reduce the amount of packaging and who use packaging made from recycled materials.

BUY OR BARTER USED GOODS: When possible, buy, barter, or trade used items to eliminate the need for packaging.

REFUSE: When offered excess packaging, such as plastic or paper grocery bags, refuse them. Either bring your own reusable bags or skip the bag altogether. Patronize stores that allow you to bring and fill your own containers, and skip the packaging completely.

MAKE YOUR OWN CLEANING SUPPLIES: As you have seen, you can make your own cleaning supplies quickly, easily, efficiently, and with just a few basic items. Making your own supplies not only saves you money and exposure to chemicals, it also cuts down on the amount of packaging your purchases require.

MAKE YOUR OWN PERSONAL CARE PRODUCTS: Once you are comfortable creating the products you clean your home with, explore creating your own cosmetics. Many recipes can be found online, and you can trust that you are using fewer toxins when you create products from your own natural supplies.

EASY WAYS TO REDUCE PAPER CONSUMPTION

Reducing paper consumption is one of the easiest green changes you can make, and the effects can readily be seen in your wallet and in the environment. Switching many of your disposable paper products is easy and cheap (or free!). If you're using cloth instead of paper for personal care and cleaning, the additional laundry will be negligible. You'll save hundreds of dollars a year on your household budget, and you'll reduce your family's contributions to our country's landfills and chemical emissions. Switching to reusable products is not the only way to cut down on your paper consumption. There are also some simple ways to cut back on paper use in your home.

The average American generates 4.5 pounds of trash every day.

MAIL: The paper industry cuts down about 1.5 trees per household every year to make all of that unwanted junk mail piling up in your mailbox.[307] Opting out of advertising mail, switching to electronic billing for your financial and personal accounts, and ending delivery of telephone books is good for the environment and saves you the time and energy of sorting through it all.

ENTERTAINMENT: Switch your entertainment consumption to electronic media. Every year, we send 2 million books, 350 million magazines, and 24 billion newspapers to the landfill.[308] Reduce your own footprint: Switch to electronic subscriptions for your magazines and newspapers, use the library to borrow books, and download music CDs online. If you're going out to a movie, concert, or sports event, check on whether you can bring your own snacks instead of buying individual, prepackaged portions while you're there.

SHOPPING: Reusable bags are no longer revolutionary green news, and they are not the only way you can green your shopping trips. Rather than wasting paper on grocery lists, look into smartphone apps that let you create and share lists with family members. Search for a store that will let you buy dry goods, produce, and personal care products in bulk using your own containers. When you do purchase goods in containers, find unique ways to reuse the containers. When paying, decline a printed receipt or use electronic receipt options where available.

WORK AND SCHOOL: Work and collaborate using electronic methods where possible: Use an intranet or online collaboration system, or share documents using a thumb drive. If you must print documents, narrow the margins, print on both sides of the paper, print only what is necessary, and use environmentally friendly paper (100 percent recycled, acid- and chlorine-free). Use paper from the recycling bin for notepads.

NAPKINS: Americans use an average of 6 paper napkins each day, or around 2,200 paper napkins every year.[309] If you stacked up all the napkins that Americans use every year, the pile would be greater than 1.7 billion feet — more than 330,000 miles. To put that into perspective, the moon is only 238,857 miles away.

Try using linen or hemp napkins instead. You can even repurpose old cotton T-shirts — cut them to the size you want, and then either serge around the outside or turn and top-stitch them.

ON THE GO: CONVENIENCE AWAY FROM HOME

Portable does not have to mean disposable! There are many easy solutions for living on the go; it just takes a little bit of planning the first time. After that, it is more hassle to buy the disposable stuff than it is to reuse what you already own. Check out these convenient alternatives for staying green away from home.

BABY WIPES

The easiest way to bring reusable wipes along is to pack wipes and a small spray bottle with wipe solution. Simply spray each wipe as needed. For the ultimate in travel convenience, get an old travel wipe container or other container, and add water or wipe solution directly to your wipes. Remember to empty the wipe container frequently, as leaving the solution in there can cause mildew. For wipe solution recipes, see the "Baby Wipes" section in the "For Parents" chapter (page 239).

NAPKINS

Instead of grabbing paper napkins the next time you eat out, try keeping a couple of cloth napkins in your backpack or purse. You can often use cloth napkins a few times before you need to wash them. Many of us grab a few extra paper napkins from fast food restaurants to keep in the car. Instead, store some of your repurposed T-shirt napkins (one- or two-ply) in the car.

SNACK CONTAINERS

Rather than employing single-use plastic snack bags, look for reusable options. Buy cloth snack and sandwich bags, or make them yourself. You can find bags that are lined either with polyurethane-laminated fabric (PUL) (make certain the label says "food safe") or another water-resistant fabric and that conveniently close with zippers or Velcro. Rigid options include BPA-safe plastic containers, glass containers with locking lids, and even jars.

DRINK CONTAINERS

Invest in reusable water bottles. They come in a variety of options, from stainless steel water bottles that are not lined with plastic (which can contain BPA)[310] to glass to even lightweight, BPA-free plastic versions. Keep a few to use at home, at work, and on the go.

DINNERWARE (PLATES, CUPS, AND CUTLERY)

Keep a picnic bag or basket stocked with lightweight enamelware or stainless steel plates and cutlery. You can find them in the camping section. If you must use disposable dinnerware, look for brands that are compostable and are made from more sustainable materials, such as bamboo, wheatgrass, sugarcane, or potato starch.

HAND SANITIZER

> **Make your own all-natural hand sanitizer using essential oils for when you are out and about.**

There are a host of antibacterial hand sanitizers on the market. With few exceptions, correct hand washing is as effective at stopping the spread of germs as using hand sanitizer. But if you're looking for a natural hand sanitizer for those times when hand washing isn't possible, here are a couple of recipes. If you make the recipes without aloe vera, use them to wipe down shopping cart handles, spray public toilet seats, or clean off picnic tables or other surfaces where your family will be eating or lounging. With aloe vera gel, the recipes make bacteria-fighting gel hand sanitizers.

THIEVES OIL BLEND SPRAY

25 drops Thieves Oil Blend (page 58)
Water

Fill a two-ounce spray container about half full with water. Add the Thieves Oil Blend. Shake well before use.

THIEVES OIL BLEND HAND SANITIZER

25 drops Thieves Oil Blend (page 58)
2 tablespoons aloe vera gel
Water

For a gel hand sanitizer, mix the Thieves Oil Blend with the aloe vera gel in a two-ounce squeezable container, then fill to the top with water. Shake well before use.

ESSENTIAL HAND SANITIZING SPRAY OR GEL

30 drops essential oils (choose from among cinnamon, clove bud, eucalyptus, lemon, lemon balm, ravensara, rosemary, and thyme)
Water or aloe vera

Follow the directions for either the Thieves Oil Blend Spray or the Thieves Oil Blend Hand Sanitizer, replacing the Thieves Oil Blend ingredients with your combination of essential oils.

A FEW MORE TIPS FOR TRAVELING

Here are a few more ways you can reduce your environmental impact when traveling.

PACK SNACKS AND MEALS INSTEAD OF EATING OUT. Bringing your own food saves money and eliminates all of the packaging used with fast food. When you take a longer road trip, pack a cooler and shop at farmers markets and grocery stores along the way. Buying local is a wonderful way to experience the flavor of the area.

If you do eat out, try going to restaurants that provide re-usable dinnerware. Many restaurants are utilizing more environmentally friendly options at lower price ranges for families who prefer these options.

REUSE CONTAINERS. Rather than buying travel-size personal care products, buy reusable containers and refill them before you leave. When flying, make certain you check any applicable size restrictions. In addition, when you travel by airplane, don't throw out the clear, quart-sized bags you use for your carry-on bag. Keep them in your bag for the next trip.

RENT A HOUSE OR APARTMENT. Consider renting a house or apartment rather than staying at a hotel. Besides likely saving you money, you will be able to cook your food, saving more money and avoiding disposable products.

USE PUBLIC TRANSPORTATION. Take a bus or the subway, and choose a smaller vehicle if you need to rent one. Look for hybrid versions. Walk to local areas to save energy and get a better feel for your new location.

TREAD LIGHTLY. Whether participating in tourist attractions or hiking through national parks, consider how your actions affect the world around you.

DON'T FORGET ABOUT REDUCING YOUR ENERGY IMPACT. Prepare your home for your departure by lowering (or raising, depending on the time of year) the thermostat, lowering the temperature on the hot water heater, and unplugging items that won't be used. Use timers for any lights or other items that need to be left on.

REDUCING YOUR ENERGY CONSUMPTION

Cutting back on what you use does not only apply to material goods. Remember that production processes have a major impact on the environment and hence your home. Reducing energy consumption is another way to reduce the impact of chemicals in your life.

There are many simple things you can do around your house to reduce your energy consumption and reduce the resources it takes to run your home. Not only will you help the environment, you will also help yourself save money.

UNPLUG: Not using an item? Unplug it. Even in standby mode, your electrical items are drawing electricity. If you have a charging station or an entertainment system, consider using surge protectors. When not in use, just flip the off switch to easily turn off power to all the items.

> Unplug items to conserve energy and save money.

ENERGY-EFFICIENT APPLIANCES: When replacing items, look for labels that state the item is energy efficient. In 1992, the EPA and the Department of Energy worked together to create Energy Star. Items carrying the Energy Star service mark use 20–30 percent less energy than the amount determined acceptable by federal standards.[311] The percentage varies based on the product. Begun as a voluntary labeling program for computer and printer products, the Energy Star program quickly expanded to include other residential products.

ECO-FRIENDLY ENERGY: Many electric companies now offer power from renewable sources instead of fossil fuels. Research companies in your area to see which offer renewable energy. Other companies offer credits for homeowners and businesses that have solar panels or wind turbines plugged into the grid. With such programs, many consumers can offset the cost of any excess electricity they use during certain parts of the year.

INSULATE YOUR HOME: As much as 30 percent of your home's heat escapes through the roof. Insulating your attic is the most cost-effective way to save energy. Cavity-wall insulation can also save a significant amount of heat loss. However, the greatest amount of heat is lost through drafts. Check doors and windows for any drafts. There are many products on the market to stop drafts and lower your heating bill. Does your home suffer in the heat? Consider adding a radiant barrier to reduce the amount of radiant heat in your attic and thus lower the temperature in your home.

LOWER THE THERMOSTAT: Lowering your thermostat by 1 degree during winter can reduce your energy consumption by 5 percent. Consider using a timer to control your thermostat. Set the temperature lower for the times when you aren't home or don't need your home as warm. Do the opposite during the summer, when you are running the air conditioner.

SAVE THE WATER: Quite a bit of energy is lost when unused water runs down your drain. Turn the water off when you aren't using it, such as when you're brushing your teeth or shaving. If you are waiting for running water to heat up, consider using the water instead of wasting it. Put a container underneath the faucet, and use the water on your plants.

TURN OFF LIGHTS: Leaving lights on in empty rooms wastes more energy than you might realize. Turn the lights off when you leave a room. Use timers on lights that need to be turned on or off on a regular schedule, such as lights that turn on so that you don't arrive at a dark house. Consider using motion sensors for outdoor lights or in rooms where lights are often left on. Motion sensors can be set to different amounts of time, and are often an asset in homes with small children.

ALLOW COMPUTERS TO HIBERNATE: Turning off your computer or handheld device each time you get up from your desk might not be practical. Use a sleep mode or hibernate feature to conserve energy during periods of inactivity. Your computer or handheld will be ready to go as soon as you need it.

RECYCLE

Lastly, we come to recycling. Recycling involves changing waste materials into new, useful materials that can then be used to make consumer goods. By recycling items, we use fewer raw materials, reduce the amount of energy it takes to turn raw materials into products, lower greenhouse gas emissions, reduce air and water pollution, and reduce the need for waste disposal. Recycled materials can include glass, paper, metal, plastic, textiles, and electronics.

Although the recycling movement began to take hold in the 1980s, recycling has a long history. Archaeologists studying waste areas from early cultures have noted that during times of scarce supplies, fewer items made it into waste areas. This is due in part to reduced consumption but is also due to the fact that more items were recycled during those times.

> More than 75 percent of solid waste is recyclable, but we recycle only a fraction of that.

Drop-off recycling centers exist in many cities. More and more cities are providing curbside recycling for residences and businesses, which makes it even easier to recycle.

TIPS FOR RECYCLING

GET INFORMED: Find out what your recycling center accepts, what items can be picked up curbside, and what items need to be dropped off. Some recycling centers have limited drop-off hours, so find out what times are available.

DETERMINE HOW MUCH SPACE YOU CAN REASONABLY DEVOTE TO RECYCLING: Making a trip to the recycling center every day doesn't make sense, but neither does keeping items that clutter up your home, are a fire hazard, and end up being more than you can take in one trip. Balance the number of trips you make to the recycling center against the space that the items to be recycled take up in your home.

SEPARATE YOUR RECYCLABLES: Use different bins to separate recyclable materials. It is easier to toss an item into the proper bin than to sort through all the items later. By sorting as you go, you will have everything ready to put in its proper location at your local drop-off center.

WASH YOUR RECYCLABLES: Rinsing a glass jar is much easier to do when the mess is fresh, and you won't be inviting pests to come raid your recycling and your home.

CONCLUSION

Taking simple steps to achieve a less toxic home means implementing small changes, one at a time. Whether you decide to start by replacing some of your cleaning products with all-natural versions, by adding reusable products to avoid toxins in disposable products, or by adopting one of the many other suggestions in this book, we hope you will feel confident to take a step or two in a healthier direction.

When you do start taking these simple steps, you will also be sending a message to the companies making the products that are harming our health and our environment. The easiest way to turn the toxic tide of household cleaning products is to stop buying them. The average American household spends more than $600 on cleaning products every year.[312] If we stop buying toxic products, they will stop making them.

You do not need to overhaul your entire life in one day. By using ingredients you already have at home and replacing one commercial product at a time, you can make many easy, all-natural recipes that will mean a healthier you.

ENDNOTES

[1] IFRA (International Fragrance Association). 2010. Ingredients. http://www.ifraorg.org/en-us/Ingredients_2.

[2] Welsh, M. S., Lamesse, M., Karpinski, E. (2000). "The verification of hazardous ingredients disclosures in selected material safety data sheets." *Appl Occup Environ Hyg.* 15(5):409-20.

[3] Acharya, P. V. Narasimh. (April 1997). "Irreparable DNA-Damage by Industrial Pollutants in Pre-mature Aging, Chemical Carcinogenesis and Cardiac Hypertrophy: Experiments and Theory." 1st International Meeting of Heads of Clinical Biochemistry Laboratories, Jerusalem, Israel. Work conducted at Industrial Safety Institute and Behavioral Cybernetics Laboratory, University of Wisconsin, Madison.

[4] Thomson/Gale and Jacqueline L. Longe. *The Gale Encyclopedia of Cancer: A Guide to Cancer and Its Treatments*. Second Edition. Farmington Hills, MI: Gale, 2005: 137.

[5] Guo, Y. L., Lambert, G. H., Hsu, C. C. (September 1995). "Growth abnormalities in the population exposed in utero and early postnatally to polychlorinated biphenyls and dibenzofurans." *Environ Health Perspect.* 103(6):117-22.

[6] "National Report on Human Exposure to Environmental Chemicals." Centers for Disease Control and Prevention, Department of Health and Human Services. http://www.cdc.gov/exposurereport.

[7] "Green Remediation." U.S. Environmental Protection Agency.

[8] Field, J. A., Sierra-Alvarez, R. (September 2008). "Microbial transformation and degradation of polychlorinated biphenyls." *Environ Pollut.* 155(1):1-12.

[9] "Case Studies of Green Remediation: Re-Solve, Inc., North Dartmouth, MA." Green Remediation. U.S. Environmental Protection Agency. http://www.clu-in.org/greenremediation/subtab_d29.cfm

[10] Olney, J. W. (2002). "New insights and new issues in developmental neurotoxicology." *Neurotoxicology.* 23(6):659-68.

[11] Nadler, J. V., Bruce, W. P., and Carl, W. C. (1978). "Intraventricular kainic acid preferentially destroys hippocampal pyramidal cells." *Nature*, 271(5646):676-77.

[12] Jevtovic-Todorovic, V., Hartman, R. E. et al. (2003). "Early exposure to common anesthetic agents causes widespread neurodegeneration in the developing rat brain and persistent learning deficits." *J Neurosci.* 23(3):876-82.

[13] Hodge, A. Trevor. *Roman Aqueducts and Water Supply.* London: Duckworth, 2002.

[14] Adams, M. E., Olivera, B. M. (1994) "Neurotoxins: overview of an emerging research technology." *Trends Neurosci.* 17(4):151-55.

[15] Nazaroff, W. W., Coleman, B. K., et al. (2006). "Indoor Air Chemistry: Cleaning Agents, Ozone and Toxic Air Contaminants." Prepared for the California Air Resources Board and the California Environmental Protection Agency: California Air Resources Board. http://www.arb.ca.gov.

[16] Bello, A., Quinn, M. M., et al. (November 2010). "Quantitative assessment of airborne exposures generated during common cleaning tasks: a pilot study." *Environ Health.* 9:76.

[17] "An Introduction to Indoor Air Quality (IAQ): Volatile Organic Compounds (VOCs)." http://www.epa.gov.

[18] Orange, E. (September/October 2010) "From eco-friendly to eco-intelligent." *Futurist.* 44(5):28.
Dahl, R. (June 2010). "Green washing: do you know what you're buying?" *Environ Health Perspect.* 118(6):A246-52.

[19] Orange, supra.

[20] Ashkin, Stephen, and David Holly. *Green Cleaning for Dummies: ISSA Special Edition.* Hoboken, NJ: Wiley, 2007.

[21] Green Marketing Exposed. http://www.marketingdegree.net/green-marketing-exposed. See also *TerraChoice, The Sins of Greenwashing Home and Family Edition 2010: A Report on Environmental Claims Made in the North American Consumer Market*, available at http://www.sinsofgreenwashing.org/index35c6.pdf.

[22] TerraChoice. The 73 percent increase occurred from the 2009 to the 2010 TerraChoice study.

[23] "EWG Cleaners Database Hall of Shame." Retrieved November 1, 2013 from http://static.ewg.org/reports/2012/cleaners_hallofshame/cleaners_hallofshame.pdf.

[24] Glaser, A. (2004). "The ubiquitous triclosan: a common antibacterial agent exposed." Beyond Pesticides/National Coalition Against the Misuse of Pesticides. 24(3).

[25] Barnett, Sloan. *Green Goes with Everything: Simple Steps to a Healthier Life and a Cleaner Planet*. New York: Atria Books, 2008.

[26] 15 U.S.C §§ 2601-97

[27] Reuben, S. "Reducing Environmental Cancer Risk: What We Can Do Now." U.S. Department of Health and Human Services 2008–2009 Annual Report, 22.

[28] Lucier, G. W., Schecter, A. (October 1998). "Human exposure assessment and the national toxicology program." *Environ Health Perspect*. 106(10).

[29] "Guides for the Use of Environmental Marketing Claims." 16 C.F.R. § 260.1(a). (1998). http://www.ftc.gov/os/2012/10/greenguides.pdf.

[30] "EWG's 2012 Guide to Healthy Cleaning." FAQ. Retrieved November 1, 2013 from http://www.ewg.org/guides/cleaners/content/faq.

[31] Thomas, Patricia. *What's in This Stuff?: The Hidden Toxins in Everyday Products and What You Can Do About Them*. New York: Penguin Group, 2008: 102.

[32] Entani, E., Asai, M., et al. (August 1998). "Antibacterial action of vinegar against food-borne pathogenic bacteria including Escherichia coli O157:H7." *J Food Prot*. 61(8):953-59.

[33] Li, L., Valenzuela-Martinez, C., et al. (November 2012). "Inhibition of clostridium perfringens spore germination and outgrowth by lemon juice and vinegar product in reduced NaCl roast beef." *J Food Sci*. 77(11):M598-603.

[34] "Borax." PAN Pesticide Database. Retrieved November 1, 2013 from http://www.pesticideinfo.org.

[35] Hill, C. N. *A Vertical Empire: The History of the UK Rocket and Space Programme, 1950–1971*. London: Imperial College Press, 2001.

[36] Schep, L. J., Slaughter, R. J., et al. (September 2009). "A seaman with blindness and confusion." *BMJ* 339: b3929.

[37] Brent, J. (May 2009). "Fomepizole for ethylene glycol and methanol poisoning." *N Engl J Med*. 360(21):2216-23.

[38] Schep, L. J., Slaughter, R. J., et al. (July 2009). "Diethylene glycol poisoning." *Clin Toxicol (Phila).* 47(6):525-35.

[39] Worwood, Valerie A. *The Complete Book of Essential Oils and Aromatherapy: Over 600 Natural, Non-Toxic and Fragrant Recipes to Create Health – Beauty – a Safe Home Environment.* Novato, CA: New World Library, 1991.

[40] Warnke, P. H., Becker, S. T., et al. (2009). "The battle against multi-resistant strains: Renaissance of antimicrobial essential oils as a promising force to fight hospital-acquired infections." *J Cranio Maxill Surg.* 37(7):392-97.

[41] Worwood, supra.

[42] Finsterer, J. (April 2002). "Earl grey tea intoxication." *Lancet.* 359(9316):1484.

[43] American Chemical Society (August 28, 2001). "Catnip Repels Mosquitoes More Effectively Than DEET." *ScienceDaily.* Retrieved October 31, 2013 from http://www.sciencedaily.com/releases/2001/08/010828075659.htm.

[44] Srivastava, J. K., Shankar, E., Gupta, S. (November 2010). "Chamomile: A herbal medicine of the past with bright future." *Mol Med Rep.* 3(6):895-901.

[45] Tayel, A. A., El-Tras, W. F. (2009). "Possibility of fighting food borne bacteria by Egyptian folk medicinal herbs and spices extracts." *J Egypt Public Health Assoc.* 84(1-2):21-32.

[46] Premanathan, M., Rajendran, S., et al. (September 2000). "A survey of some Indian medicinal plants for anti-human immunodeficiency virus (HIV) activity." *Indian J Med Res.* 112:73-77.

[47] Khan, A., Safdar, M., et al. (December 2003). "Cinnamon improves glucose and lipids of people with type 2 diabetes." *Diabetes Care.* 26(12):3215-18.

[48] Benencia, F., Courrèges, M. C. (2000). "In vitro and in vivo activity of eugenol on human herpesvirus." *Phytother Res.* 14(7):495-500.

[49] Orihara, Y., Hamamoto, H., et al. (January 2008). "A silkworm baculovirus model for assessing the therapeutic effects of antiviral compounds: characterization and application to the isolation of antivirals from traditional medicines." *J Gen Virol.* 89 (Pt 1):188-94.

[50] Pattanaik, S., Subramanyam, V. R., Kole, E. C. (1996). "Antibacterial and antifungal activity of ten essential oils in vitro." *Microbios.* 86(349):237-46.

[51] El-Zemity Saad, E., Hussein, R., et al. (December 2006). "Acaridicidal activities of some essential oils and their monoterpenoidal constituents against house dust mite, dermatophagoides pteronyssinus." *J Zhejiang Univ Sci B.* 7(12):957-62.

[52] Salari, M. H., Amine, G., et al. (February 2006). "Antibacterial effects of Eucalyptus globulus leaf extract on pathogenic bacteria isolated from specimens of patients with respiratory tract disorders." *Clin Microbiol Infect.* 12(2):194-96.

[53] Habanananda, T. (October 2004). "Non-pharmacological pain relief in labour." *J Med Assoc Thai.* 87(Suppl.3):S194-202.

[54] Shiina, Y., Funabashi, N., et al. (September 2008). "Relaxation effects of lavender aromatherapy improve coronary flow velocity reserve in healthy men evaluated by transthoracic Doppler echocardiography." *Int J Cardiol.* 129(2):193-97.

[55] Hongratanaworakit T. (August 2011). "Aroma-therapeutic effects of massage blended essential oils on humans." *Nat Prod Commun.* 6(8):1199-204.

[56] Salari, supra.

[57] Kucera, L. S., Cohen, R. A., Herrmann, E. C. (July 1965). "Antiviral activities of extracts of the lemon balm plant." *Ann N Y Acad Sci.* 130(1):474-82.

[58] Shadab, Q., Hanif, M., and Chaudhary, F. M. (1992). "Antifungal activity by lemongrass essential oils." *Pak J Sci Ind Res.* 35, 246-49.

[59] Verzera, A. Trozzi, G., et al. (May 2004). "Biological lemon and sweet orange essential oil composition." *Flavour Frag J.* 19(6):544-48.

[60] Faleiro, L., Miguel, G. (October 2005). "Antibacterial and antioxidant activities of essential oils isolated from thymbra capitata L. (Cav.) and Origanum vulgare L." *J Agric Food Chem.* 53(21):8162-68.

[61] Schmidt, E., Bail, S., et al. (August 2009). "Chemical composition, olfactory evaluation and antioxidant effects of essential oil from Mentha x piperita." *Nat Prod Commun.* 4(8):1107-12.

[62] Harrison, Lorraine. *RHS Latin for Gardeners: Over 3,000 Plant Names Explained and Explored.* London: Mitchell Beazley, 2012: 224.

[63] Stearn, William T. *Botanical Latin.* Portland, OR: Timber Press, 2004: 456.

[64] Hussain, A. I., Anwar, F., et al. (2010). "Seasonal variation in content, chemical composition and antimicrobial and cytotoxic activities of essential oils from four Mentha species." *J Sci Food Agr.* 90(11):1827-36.

[65] Nenoff, P., Haustein, U. F., Brandt, W. (1996). "Antifungal activity of the essential oil of Melaleuca alternifolia (tea tree oil) against pathogenic fungi in vitro." *Skin Pharmacol.* 9(6):388-94.

[66] Bishop, C. D. (1995). "Antiviral activity of the essential oil of Melaleuca alternifolia (maiden and betche) cheel (tea tree) against tobacco mosaic virus." *J Essen Oil Res.* 7(6):641-44.

[67] Pierce, Andrea. *The American Pharmaceutical Association Practical Guide to Natural Medicines.* New York: Stonesong, 1999: 338-40.

[68] Ramsewak, R. S.; Nair, M. G. et al. (April 2003). "In vitro antagonistic activity of monoterpenes and their mixtures against 'toe nail fungus' pathogens." *Phytother Res.* 17(4):376-79.

[69] Sánchez-Pérez, R., Belmonte, F. S., et al. (April 2012). "Prunasin hydrolases during fruit development in sweet and bitter almonds." *Plant Physiol.* 158(4):1916-32.

[70] Yang, D., Pornpattananangkul, D., et al. (2009). "The antimicrobial activity of liposomal lauric acids against Propionibacterium acnes." *Biomaterials.* 30(30):6035-40.

[71] Ogbolu, D. O., Oni, A. A., et al. (June 2007). "In vitro antimicrobial properties of coconut oil on Candida species in Ibadan, Nigeria." *J Med Food.* 10(2):384-87.

[72] Yang, supra.

[73] Place, A. R. (September 1992). "Comparative aspects of lipid digestion and absorption: physiological correlates of wax ester digestion." *Am J Physiol.* 263(3):R464-71.

[74] Wołosik, K., Knaś, M., et al. (January 2013). "The importance and perspective of plant-based squalene in cosmetology." *J Cosmet Sci.* 64(1):59-66.

[75] Holtzapple, M. T., Lundeen, J. E. (1992). "Pretreatment of lignocellulosic municipal solid waste by ammonia fiber explosion (AFEX)." *Appl Biochem Biotechnol.* 34-35(1):5-21.

[76] Lipka, Nate. (May 15, 2012). "Help Save 571,230,000 Pounds of Paper Towels." Retrieved October 31, 2013 from http://www.earth911.com/news/2012/05/how-to-use-one-paper-towel-and-save-paper/news/2012/05/15/how-to-use-one-paper-towel-and-save-paper

[77] Natural Resources Defense Council. "Environmental Ratings of Household Tissue Paper Products by Category." http://www.nrdc.org.

[78] McBain, A. J., Bartolo, R. G., et al. (2003). "Exposure of sink drain microcosms to triclosan: population dynamics and antimicrobial susceptibility." *Appl Environ Microbiol.* 69(9):5433-42.

[79] Aiello, A. E., Marshall, B., et al. (October 2005). "Antibacterial cleaning products and drug resistance." *Emerg Infect Dis.* 11(10):1565-70.

[80] Adolfsson-Erici, M., Pettersson, M., et al. (2002). "Triclosan, a commonly used bactericide found in human milk and in the aquatic environment in Sweden." *Chemosphere.* 46:1485-89.

[81] Curtis, V. A. (August 2007). "Dirt, disgust and disease: a natural history of hygiene." *J Epidemiol Community Health.* 61(8): 660-64.

[82] Curtis, V. A., Biran, A. (Winter 2001). "Dirt, disgust and disease: is hygiene in our genes?" *Perspect Biol Med.* 44(1):17-31.

[83] Hart, B. L. (1990). "Behavioural adaptations to pathogens and parasites: five strategies." *Neurosci Biobehav Rev.* 14(3):273-94.

[84] Barnes, David S. *The Great Stink of Paris and the Nineteenth Century Struggle Against Filth and Germs.* Baltimore, MD: Johns Hopkins University Press, 2006.

[85] Hart, supra.

[86] Kiesecker, J., Skelly, D. et al. (1999). "Behavioral reduction of infection risk." *Proc Natl Acad Sci.* 96(16):9165-68.

[87] Cipolla, Carlo M. *Before the Industrial Revolution: European Society and Economy 1000–1700.* London: W. W. Norton and Company, 1980.

[88] Johansson, Ingegard, and P. Somasundaran, eds. *Handbook for Cleaning/ Decontamination of Surfaces*, vol. 1. Amsterdam: Elsevier, 2007: 189.

[89] Johansson and Somasundaran, supra, 11.

[90] Orey, Cal. *The Healing Powers of Olive Oil: A Complete Guide to Nature's Liquid Gold.* New York: Kensington Books, 2000: 22-23.

[91] Jakob, U., Winter, J., et al. (November 2008). "Bleach activates a redox-regulated chaperone by oxidative protein unfolding." *Cell.* 135(4):691-701.

[92] "Guidelines for the Use of Sanitizers and Disinfectants in Child Care Centers." Colorado Department of Public Health and Environment. Revised 12/01. http://www.nchd.org/Downloads/Child%20Care/Sanitizers%20%20 DisinfectantsGuideance%20Document.pdf

[93] Odabasi, M. "Halogenated Volatile Organic Compounds from the Use of Chlorine-Bleach-Containing Household Products." Slide presentation, 2008.

[94] Chapdelaine, P. Anthony, Jr., MD, MSPH. (August 1998). "The Best and Cheapest Home Disinfectant Summary of Report." *Science News.* 6(154):83-85.

[95] Awad, M. I., Denggerile, A., Ohsaka, T. (2004). "Electroreduction of peroxyacetic acid at gold electrode in aqueous media." *J Electrochem Soc.* 151:E358.

[96] Cristofari-Marguand, E., Kacel, M., et al. (2007). "Asthma caused by peracetic acid-hydrogen peroxide mixture." *J Occup Health.* 49(2):155-58.

[97] Rosen, Milton, J. and Joy T. Kunjappu. *Surfactants and Interfacial Phenomena.* Hoboken, NJ: John Wiley & Sons, 2012: 403.

[98] Rusin, P., Orosz-Coughlin, B., Gerba, C. (November 1998). "Reduction of faecal coliform, coliform and heterotrophic plate count bacteria in the household kitchen and bathroom by disinfection with hypochlorite cleaners." *J App Microbiol.* 85(5):819-28.

[99] Rusin, supra.

[100] Ak, N. O., Cliver, D. O., Kaspar, C. W. (1994). "Cutting boards of plastic and wood contaminated experimentally with bacteria." *J Food Protect.* 57:16-22.

Ak, N. O., Cliver, D. O., Kaspar, C. W. (1994). "Decontamination of plastic and wooden cutting boards for kitchen use." *J Food Protect.* 57:23-30,36.

Park, P. K., Cliver, D. O. (June-July 1997). "Cutting boards up close." *Food Quality.* 3(22):57-59.

Kass, P. H., et al. (1992). "Disease determinants of sporadic salmonellosis in four Northern California counties: a case control study of older children and adults." *Ann Epidemiol.* 2:683-96.

Abrishami, S. H., Tall, B. D., et al. (May 1994). "Bacterial adherence and viability on cutting board surfaces." *J Food Safety.* 14(2):153-72.

Snyder, Peter. "The Evaluation of Wooden vs. Polyethylene Cutting Boards Using Fluorescent Powder." Hospitality Institute of Technology and Management. Retrieved November 1, 2013 from http://www.hi-tm.com/Documents2008/cutboard-eval.pdf.

[101] Cliver, D. O. (March 2006). "Cutting boards in salmonella cross-contamination." *J AOAC Int.* 89(2):538-42(5).

[102] Willing, A. (April 2001). "Lubricants based on renewable resources — an environmentally compatible alternative to mineral oil products." *Chemosphere.* 43(1):89-98.

[103] Willing, supra.

[104] Dresdner, Michael. (April 1994). "Food-safe finishes: nontoxic treatments for your woodenware." *American Woodworker.* 74-76.

[105] DiDonato, Jessica, ed. *Fine Woodworking Best Finishing Techniques.* Newton, CT: The Taunton Press, 2011: 225.

[106] Ogbolu, supra.

[107] Yang, supra.

[108] Projan, S. J., Brown-Skrobot, S., et al. (1994). "Glycerol monolaurate inhibits the production of beta-lactamase, toxic shock toxin-1, and other staphylococcal exoproteins by interfering with signal transduction." *J Bacteriol.* 176:4204-209.

[109] Hurtado de Catalfo, G. E., de Alaniz, M. J. T., Marra, C. A. (July-August 2010). "Dietary lipids modify redox homeostasis and steroidogenic status in rat testis." *Phytother Res.* 24(7-8):717-26.

[110] Tegethoff, F. Wolfgang, ed. *Calcium Carbonate: From the Cretaceous Period into the 21st Century.* Basel, Switzerland: Birkhauser Verlag, 2001: 98.

[111] U.S. General Services Administration. "Historic Preservation—Technical Procedures: Removing Organic Stains from Marble." Retrieved November 1, 2013 from http://www.gsa.gov/portal/content/111906.

[112] Wilks, S. A., Michels, H., Keevil, C. W. (December 2005). "The survival of Escherichia coli O157 on a range of metal surfaces." *Int J Food Microbiol.* 105(3):445-54.

[113] Noyce, J. O., Michels, H., Keevil, C. W. (April 2007). "Inactivation of influenza a virus on copper versus stainless steel surfaces." *Appl Environ Microbiol.* 73(8):2748-50.

[114] Bronner, Lisa. "A Word of Caution About Vinegar and Castile Soap." *Going Green with a Bronner Mom.* Retrieved November 1, 2013 from http://lisa.drbronner.com/?p=292.

[115] Mendelson, Cheryl. *Home Comforts: The Art and Science of Keeping House*. New York: Scribner, 1999: 176.

[116] Raloff, J. (September 1996). "Sponges and sinks and rags, oh my! Where microbes lurk and how to rout them." *Science News*. 150(11):172.

[117] Sharma, M., Eastridge, J., Mudd, C. (March 2009). "Effective household disinfection methods of kitchen sponges." *Food Control*. 20(3):310-13.
Tate, N. J. (May 2006). "Bacteria in household sponges: A study testing which physical methods are most effective in decontaminating kitchen sponges." *Saint Martin's University Biology Journal*, Volume 1.

[118] Richter, C. P. (March 2010). "Automatic dishwashers: efficient machines or less efficient consumer habits?" *International Journal of Consumer Studies*. 34(2):228-34.

[119] Editors of the Green Guide Magazine. *Green Guide: The Complete Reference for Consuming Wisely*. Washington, DC: National Geographic Ventures, 2008: 50-52.
Johnston, C. J., Finkelstein, J. N., et al. (1998). "Pulmonary inflammatory responses and cytokine and antioxidant mRNA levels in the lungs of young and old C57BL/6 mice after exposure to Teflon fumes." *Inhal Toxicol*. 10(10):931-53.

[120] For more on cookware alternatives, see Loux, Renée. *Easy Green Living: The Ultimate Guide to Simple, Eco-Friendly Choices for You and Your Home*. New York: Rodale, 2008: 105-19, the book we referenced several times for this section.

[121] Barr, Tracy. *Cast Iron Cooking for Dummies*. Hoboken, NJ: Wiley Publishing Inc., 2004: Chapters 3-4.

[122] Editors of the Green Guide Magazine, supra.

[123] Perl, D., Moalem, S. (2006). "Aluminum and Alzheimer's disease: a personal perspective after 25 years." *J Alzheimers Dis*. 9(3):291-300.

[124] Gerba, Charles P. Ph.D., cited in "Dr. Germ: Here a Germ, there a germ, everywhere a . . . wait." College of Agriculture and Life Sciences, University of Arizona, February 17, 2005, cited in Beard & Cerf, *Encyclopedia Paranoiaca*. Simon & Schuster, 2012: 289.

[125] Rider, Traci, Stacy Glass, and Jessica McNaughton. *Understanding Green Building Materials*. New York: W. W. Norton & Company, Inc., 2011.

[126] For more homemade floor wax recipes, see Berthold-Bond, Annie. *Better Basics for the Home: Simple Solutions for Less Toxic Living*. New York: Three Rivers Press, 1999: 64-65.

[127] Sun, N., Shen, Y., et al. (February 2009). "Diagnosis and treatment of melamine-associated urinary calculus complicated with acute renal failure in infants and young children." *Chin Med J.* 122(3).

[128] Liu, J. M., Ren, A., et al. (February 2010). "Urinary tract abnormalities in Chinese rural children who consumed melamine-contaminated dairy products: a population-based screening and follow-up study." *Can Med Assoc J.* 82(5):439-43.

[129] Eilperin, Juliet. "Harmful Teflon Chemical to Be Eliminated by 2015." *Washington Post*, January 26, 2006; p. A01.

[130] Dietert, Rodney R., and Janice Dietert. *Strategies for Protecting Your Child's Immune System: Tools for Parents and Parents-to-Be*. Singapore: World Scientific Publishing Co., 2010: 165-66.

[131] Brooks, G. Daniel, and Robert K. Bush. (2009). "Allergens and other factors important in atopic disease." In Leslie Carroll Grammer & Paul A. Greenberger. *Patterson's Allergic Diseases* (7th ed.). Baltimore, MD: Lippincott Williams & Wilkins, 2009: 73-103.

[132] McDonald L. G., Tovey E. (October 1992). "The role of water temperature and laundry procedures in reducing house dust mite populations and allergen content of bedding." *J Allergy Clin Immunol.* 90:599-608.

[133] Austin, Daniel F. *Florida Ethnobotany*. Boca Raton, FL: CRC Press, 2004: 601-603.

[134] Hostettmann, K., and A. Marston. *Saponins*. Cambridge: Cambridge University Press, 1995:3ff.

[135] Stoffels, Karen (September 2008). "Soap nut saponins create powerful natural surfactant." *Personal Care Magazine*.

[136] Austin, supra.

[137] Anderson, S. E., Franko, J., et al. (March 2013). "Exposure to triclosan augments the allergic response to ovalbumin in a mouse model of asthma." *Toxicol Sci.* 132(1):96-106.

[138] Veldhoen, N., Skirrow, R. C., et al. (December 2006). "The bactericidal agent triclosan modulates thyroid hormone-associated gene expression and disrupts postembryonic anuran development." *Aquat Toxicol.* 80(3):217-27.

[139] Clayton, E. M., Todd, M., et al. (March 2011). "The impact of bisphenol A and triclosan on immune parameters in the U.S. population, NHANES 2003-2006." *Environ Health Perspect.* 119(3):390-96.

[140] Ziegler, D. M., Ansher, S. S., et al. (April 1988). "N-Methylation: Potential mechanism for metabolic activation of carcinogenic primary arylamines." *Proc Natl Acad Sci USA.* 85:2514-17.

[141] "Formaldehyde, 2-Butoxyethanol and 1-tert-Butoxypropan-2-ol." (2006) IARC Monographs on the Evaluation of Carcinogenic Risks to Humans. Lyon, France: International Agency for Research on Cancer. 8:39-325.

[142] Heudorf, U., Mersch-Sundermann, V., Angerer, J. (October 2007). "Phthalates: toxicology and exposure." *Int J Hyg Environ Health.* 210(5):623-34.

[143] Larsen, B. K., Bjornstad, A., et al. (June 2006). "Comparison of protein expression in plasma from nonylphenol and bisphenol A-exposed Atlantic cod (Gadus morhua) and turbot (Scophthalmus maximus) by use of SELDI-TOF." *Aquat Toxicol.* 78(Suppl.1):S25-33.

[144] Soares, A., Guieysse B., et al. (2008). "Nonylphenol in the environment: A critical review on occurrence, fate, toxicity and treatment in wastewaters." *Environ Internat.* 34:1033-49.

[145] Ward, M. P., Armstrong, R. T. (November 1998). "Pesticide use and residues on Queensland wool." *Aust Vet J.* 76(11):739-42.

[146] Preparation of Australian Wool Clips, Code of Practice 2010–2012, Australian Wool Exchange (AWEX), 2010.

[147] Barber, E. J. W. *Prehistoric Textiles: The Development of Cloth in the Neolithic and Bronze Ages with Special Reference to the Aegean.* Princeton, NJ: Princeton University Press, 1992: 31.

[148] Zaisheng, C., Ypiping, Q. (January 2003). "Using an aqueous epoxide in bombyx mori silk fabric finishing." *Text Res J.* 73:42-46.

[149] Barber, supra.

[150] Adamson, Melitta Weiss, ed. *Regional Cuisines of Medieval Europe: A Book of Essays.* New York: Rutledge, 2002: 98, 166.

[151] Cherrett, Nia and John Barrett, et al. "Ecological Footprint and Water Analysis of Cotton, Hemp, and Polyester." Stockholm Environment Institute, 2005.

[152] Tamburlini, G. (2002). "Children's health and environment: review of the evidence." European Environment Agency and WHO, Regional Office.

[153] Papinchak, H. L. (2009). "Effectiveness of houseplants in reducing the indoor air pollutant ozone." *HortTechnology*. 19:286-90.

[154] Yang, D. S. C. (August 2009). "Screening indoor plants for volatile organic pollutant efficiency." *HortScience*. 44:1377-81.

[155] Adgate, J. E. (2004). "Outdoor, indoor, and personal exposure to VOCs." *Environ Health Perspect*. 112:1386-92.

[156] Volatile Organic Compounds (VOCs) include propane, butane, ethanol, benzene, formaldehyde, acetone, and methanol.

[157] Tillett, T. (2012). "Hearts over time: cardiovascular mortality risk linked to long-term PM2.5 exposure." *Environ Health Perspect*. doi: 10.1289/ehp.120-a205a.

[158] Gomez-Serrano, V., Piriz-Almeida, F., et al. (1999). "Formation of oxygen structures by air activation. A study by FT-IR spectroscopy." *Carbon*. 37(10):1517-28.

[159] Activated Carbon, including AquaCarb (except NoRise family), VOCarb Series (except LoRise family and HgFree family), AC Series, VC Series, BevCarb Series, and UtraCarb Series. Siemens Material Safety Data Sheet. April 2011. Siemens Industry, Inc. Water Technologies Business Unit.

[160] Weshler, C. J. (2006). "Ozone's impact on public health: contributions from indoor exposures to ozone and products of ozone-initiated chemistry." *Environ Health Perspect*. 114(10):1489-96.

[161] Smith, L. L. (2004). "Oxygen, oxysterols, ouabain, and ozone: a cautionary tale." *Free Radic Biol Med*. 37(3):318-24

[162] Nicole Folchetti, ed. "22." *Chemistry: The Central Science* (9th ed.). Pearson Education, 2003: 882-83.

[163] Wilson, E. K. (March 2009). "Ozone's health impact." *Chemical & Engineering News*. 87(11):9.0.

[164] Dunston, N. C., Spivak, S. M. (1997). "A preliminary investigation of the effects of ozone on post-fire volatile organic compounds." *J Appl Fire Sci*. 6(3):231-42.

[165] Wolverton, B. C., Douglas, W. L., et al. (1989). "Interior landscape plants for indoor air pollution abatement." National Aeronautics and Space Administration report.

[166] Son K. C., Lee, S. H. (2000). "Effects of foliage plants and potting soil on absorption and adsorption of indoor air pollutants." *J Korean Soc Hort Sci.* 41:305-10.

[167] Giese, M., Bauer-Doranth, U., et al. (1994). "Detoxification of formaldehyde by the spider plant (Chlorophytum comosum)." *Plant Physiol.* 104: 1301-309.

[168] Wolverton, B. C., Wolverton, J. D. (1996). "Interior plants: their influence on airborne microbes inside energy-efficient buildings," *J Miss Acad Sci.* 41(2):99-105.

[169] Lohr, V. I., et al. (1996). "Interior plants may improve worker productivity and reduce stress in a windowless environment." *J Environ Hort.* 14:97-100.

[170] Lohr, V. I. (1996). "Particulate matter accumulation on horizontal surfaces in interiors: influence of foliage plants," *Atmos Environ.* 30:2565-68.

[171] Wolverton and Wolverton, supra.

[172] Boudreau, M. D., Beland, F. A. (2006). "An evaluation of the biological and toxicological properties of aloe barbadensis (miller), aloe vera." *J Environ Sci Health C Environ Carcinoa Ecotoxical Rev.* 24(1):103-54.

[173] Vogler, B. K., Ernst, E. (October 1999). "Aloe vera: a systematic review of its clinical effectiveness." *Br J Gen Pract.* 49(447):823-28.

[174] Wolverton, B. C., McDonald, R. C., Watkins, E. A. Jr. "Foliage plants for removing indoor air pollutants from energy-efficient homes." *Econ Bot.* 38:224-28.

[175] Wolverton, McDonald, and Watkins, supra.

[176] Huxley, Anthony, and Mark Griffiths, eds. *The New RHS Dictionary of Gardening.* Macmillan, 1992.

[177] Wolverton, McDonald, and Watkins, supra.

[178] Wolverton, McDonald, and Watkins, supra.

[179] Wolverton, B. C. *How to Grow Fresh Air*, New York: Penguin Books, 1997.

[180] Sawada, A., Oyabu, T. (January 2008). "Purification characteristics of pothos for airborne chemicals in growing conditions and its evaluation." *Atmos Environ.* 42(3):594-602.

[181] Gonçalves, E. G., Mayo, S. J. (2000). "Philodendron venustifoliatum (araceae): a new species from Brazil." Kew Bulletin (Springer). 55(2):483-86.

[182] Wolverton, McDonald, and Watkins, supra.

[183] Wolverton, McDonald, and Watkins, supra.

[184] Wolverton, McDonald, and Watkins, supra.

[185] Kim, K. J., Kil, M. J., et al. (July 2008). "Efficiency of volatile formaldehyde removal by indoor plants: contribution of aerial plant parts versus the root zone." *J Amer Soc Horti Sci.* 133 (4):521-26.

[186] Johnson, E. (November 2009). "Charcoal versus LPG grilling: a carbon-footprint comparison." *Environmental Impact Assessment Review.* 29(6):370-78.

[187] Raloff, J. (July 1991). "Cholesterol—Up in Smoke." *Science News.*

[188] Raloff, supra.

[189] Richards, S. L., Anderson, S. L., Smartt, C. T. (October 2009). "The female mosquito's quest for blood: implications for disease cycles." *Entomology and Nematology.* Florida Cooperative Extension Service ENY-855 (IN811).

[190] Burgdorfer, W., Barbour, A. G., et al. (1982). "Lyme disease—a tick-borne spirochetosis?" *Science.* 216:1317-19.

[191] Barbour, A. G., Fish, D. (1993). "The biological and social phenomenon of Lyme disease." *Science.* 260:1610-16.

[192] National Pesticide Telecommunication Network (NPTN). DEET General Fact Sheet. December. Corvallis, OR: Oregon State University, 2000.

[193] Diethyltoluamide (DEET) Chemical Survey. U.S. Environmental Protection Agency, Toxicity and Exposure Assessment for Children's Health.

[194] Heick, H. M., et al. (1988). "Insect repellent, N,N-diethyl-m-toluamide, effect on ammonia metabolism." *Pediatrics.* 82(3):373-76.

[195] Oransky, S., et al. (1989). "Seizures temporarily associated with use of DEET insect repellent-New York and Connecticut." *Morbidity and Mortality Weekly Report.* 38(39):678-80.

[196] Lipscomb, J. W., et al. (1992). "Seizure following brief exposure to the insect repellent N,N-diethyl-mtoluamide." *Ann Emerg Med.* 21(3):315-17.

[197] Roland, E. H., et al. (1985). "Toxic encephalopathy in a child after brief exposure to insect repellents." *Can Med Assoc J.* 132(2):155-56.

[198] Briassoulis, G., et al. (2001). "Toxic encephalopathy associated with use of DEET insect repellents: a case analysis of its toxicity in children." *Hum Exp Toxicol.* 20(1):8-14.

[199] Heick, H. M., et al. (1980). "Reye-like syndrome associated with use of insect repellent in a presumed heterozygote for ornithine carbamoyl transferase deficiency." *J Pediatr.* 97(3):471-73.

[200] Lipscomb, supra.

[201] Roland, supra.

[202] Public Health Agency of Canada. (2003). "Safety Tips on Using Personal Insect Repellents."

[203] Brown, M., Hebert, A. A. (1997). "Insect repellents: an overview." *J Am Acad Dermatol.* 36(2):243-49.

[204] National Institutes of Health (NIH). 2002. DEET. Hazardous Substance Database.

[205] Abou-Donia, M., et al. (2001). "Neurotoxicity resulting from coexposure to pyridostigmine bromide, DEET, and permethrin." *J Toxicol Environ Health.* (64):373-84.

[206] Abou-Donia, M., Abdel-Pahman, A., et al. (2001). "Subchronic dermal application of N,N-diethyl m-toluamide (DEET) and permethrin to adult rats, alone or in combination, causes diffuse neuronal cell death and cyto." *Exp Neurol.* 172(1):153-71.

[207] Helson, B. V. (1992). "Naturally derived insecticides: prospects for forestry use." *Forestry Chronicle.* 68:349-54.

[208] Schmutterer, Heinrich. *The Neem Tree: Sources of Unique Natural Products for Integrated Pest Management, and Medicinal, Industrial and Other Purposes.* New York: Wiley-Blackwell, 1995.

[209] Puri, H. S. *Neem: The Divine Tree* Azadirachta indica. Amsterdam: Harwood Academic Publications, 1999.

[210] Sharma S. K., Dua V. K., Sharma V. P. (March 1995). "Field studies on the mosquito repellent action of neem oil." *Southeast Asian J Trop Med Public Health.* 26(1):180-82.

[211] Dua, V. K., Nagpal, B. N., Sharma, V. P. (June 1995). "Repellent action of neem cream against mosquitoes." *Indian J Malariol*. 31(2):47-53.

[212] Sharma, supra.

[213] Sharma, V. P., Ansari, M. A., Razdan, R. K. (September 1993). "Mosquito repellent action of neem (Azadirachta indica) oil." *J Am Mosq Control Assoc*. 9(3):359-60.

[214] Ghandi, M., Lal, R., et al. (May-June 1988). "Acute toxicity study of the oil from azadirachta indica seed (neem oil)." *J Ethnopharmacol*. 23(1):39-51.

[215] Chaudhari, P. S., Vangam, S. S., et al. (2013). "Herbal plants as an ant repellent." *Internat J Bioassays*. 2(6).

[216] Peterson, C. J., et al. (2002). "Behavioral activity of catnip (lamiaceae) essential oil components to the German cockroach (blattodea: blattellidae)." *J Econ Entomol*. 95(2):377-80.

[217] Jang, Y. S., et al. (2005). "Vapor phase toxicity of marjoram oil compounds and their related monoterpenoids to blattella germanica (orthoptera: blattellidae)." *J Agric Food Chem*. 53(20):7892-98.

[218] Zhu, J. J., Zeng, X. P., et al. (2009). "Efficacy and safety of catnip (Nepeta cataria) as a novel filth fly repellent." *Med Veterin Entomol*. 23:209-16.

[219] "Beneficial insect habitat in an apple orchard: Effects on pests." Research Brief #71. Center for Integrated Agricultural Systems, College of Agricultural and Life Sciences, University of Wisconsin-Madison. September 2004.

[220] New, T. R. (2002). "Prospects for extending the use of Australian lacewings in biological control." *Acta Zool Acad Sci Hung*. 48(Suppl.2):209-16.

[221] Prete, Fredrick R. *The Praying Mantids*. Baltimore, MD: Johns Hopkins University, 1999: 27-29, 101-103.

[222] Dube, S., Upadhyay, P. D., Tripathi, S. C. (1989). "Antifungal, physicochemical, and insect-repelling activity of the essential oil of Ocimum basilicum." *Can J Bot*. 67(7): 2085-87

[223] Harley, R. M., Atkins, S., Budantsev, A. L. et al. (2004). "Labiatae" pages 167-275. In Kubitzki, Klaus (editor) and Kadereit, Joachim W. (volume editor). *The Families and Genera of Vascular Plants* volume VII. Springer-Verlag: Berlin, Germany, 2004.

[224] Whitten, W. M. (March 1981). "Pollination Ecology of monarda didyma, M. clinopodia, and hybrids (lamiaceae) in the Southern Appalachian Mountains." *Am J Bot*. 68(3):435-42.

[225] American Chemical Society (August 2001). "Catnip Repels Mosquitoes More Effectively Than DEET." *Science Daily*.

[226] "Termites Repelled by Catnip Oil." Southern Research Station, U.S. Department of Agriculture—Forest Service. March 26, 2003.

[227] Schultz, G., Peterson, C., Coats, J. (May 25, 2006). "Natural insect repellents: activity against mosquitoes and cockroaches." In Rimando, Agnes M.; Duke, Stephen O. Natural Products for Pest Management," American Chemical Society Symposium Series.

[228] Metcalf, Robert L. "Insect Control" in *Ullmann's Encyclopedia of Industrial Chemistry*. Wiley-VCH, Weinheim: 2002.

[229] Kim, J. K., Kang, C. S., et al. (2005). "Evaluation of repellency effect of two natural aroma mosquito repellent compounds, citronella and citronellal." *Entomological Research*. 35(2):117-20.

[230] Santos, Robert L. (1997). "Section Three: Problems, Cares, Economics, and Species." *The Eucalyptus of California*. California State University.

[231] Jaenson, T. G., et al. (July 2006). "Repellency of oils of lemon eucalyptus, geranium, and lavender and the mosquito repellent mygga natural to ixodes ricinus (acari: ixodidae) in the laboratory and field." *J Med Entomol*. 43(4):731-36.

[232] Carroll, S. P., Loye, J. (2006). "A commercially available botanical insect repellent as effective as DEET." *J Amer Mosquito Control Association*. 22(3):507-14.

[233] Anwar, A., Groom, M., Sadler-Bridge, D. (June 2009). "Garlic: from nature's ancient food to nematicide." *Pesticide News*. 84:18-20.

[234] Jain, N., Aggarwal, K. K., et al. (2001). "Essential oil composition of geranium (Pelargonium sp.) from the plains of Northern India." *Flavour Frag J*. 16:44-46.

[235] Rahman, G. K. M. M., Motoyama, N. (1998). "Identification of the active components of garlic causing repellent effect against the rice weevil and the wheat flour beetle." *Nihon Oyo Doubutsu Konchu Gakkai Taikai Koen Youshi*. 42:211.

[236] Jaenson, supra.

[237] Worwood, supra.

[238] Jaenson, supra.

[239] Kim, supra.

[240] Oyedele, A. O., et al. (2002). "Formulation of an effective mosquito-repellent topical product from lemongrass oil." *Phytomedicine*. 9(3):259-62.

[241] Simon, James E., Alena F. Chadwick, and Lyle E. Craker. *Herbs: An Indexed Bibliography 1971-1980 the Scientific Literature on Selected Herbs, and Aromatic and Medicinal Plants of the Temperate Zone*. Hamden, CT: Archon Books, 1984: 770.

[242] Ansari, M. A., et al. (2000). "Larvicidal and mosquito repellent action of peppermint (Mentha piperita) oil." *Bioresource Technol*. 71(3):267-71.

[243] Germplasm Resources Information Network (GRIN) (September 2005). "Taxon: Matricaria discoidea DC." Taxonomy for Plants. USDA, ARS, National Genetic Resources Program, National Germplasm Resources Laboratory, Beltsville, Maryland.

[244] Momen, F. M., et al. (2001). "Repellent and oviposition-deterring activity of rosemary and sweet marjoram on the spider mites tetranychus urticae and eutetranychus orientalis (acari: tetranychidae)." *Acta Phytopatho Hun*. 36(1-2):155-64.

[245] Pojar, Jim. *Plants of the Pacific Northwest Coast: Washington, Oregon, British Columbia, and Alaska* (revised ed.). Auburn, WA: Lone Pine Publishing, 2004.

[246] Panneerselvam, C., Murugan, K., et al. (December 2012). "Mosquito larvidicidal, pupicidal, adulticidal, and repellent activity of Artemisia nilagirica against Anopheles stephensi and Aedes aegypti." *Parasitol Res*. 111(6):2241-51.

[247] Fields, P., Allen, S., et al. (July 2002). "Standardized testing for diatomaceous earth." Proceedings of the Eighth International Working Conference of Stored-Product Protection. York, U.K.: Entomological Society of Manitoba.

[248] Lartigue, E. C., Rossanigo, C. E. (2004). "Insecticide and anthelmintic assessment of diatomaceous earth in cattle." *Veterinaria Argentina*. 21(209):660-74.

[249] Fernandez, M. I., Woodward, B. W., Stromberg, B. E. (1998). "Effect of diatomaceous earth as an anthelmintic treatment on internal parasites and feedlot performance of beef steers." *Anim Sci*. 66(3):635-41.

[250] May, R. M. (1988). "How many species are there on earth?" *Science*. 241:441-1449.

[251] Cockell, Charles, Christian Koeberl, and Iain Gilmou, eds. *Biological Processes Associated with Impact Events* (1st ed.). Berlin: Springer, 2006: 197-219.

[252] Vandermeer, John H. *The Ecology of Agroecosystems*. Sudbury, MA: Jones and Bartlett, 2010.

[253] "Monsanto Pulls Roundup Advertising in New York." *Wichita Eagle*. Nov. 27, 1996.

[254] "Complaints halt herbicide spraying in Eastern Shore." *CBC News*. June 16, 2009.

[255] Quastel, J. H. (1950). "2,4-dichlorophenoxyacetic acid (2,4-d) as a selective herbicide." *Agricultural Control Chemicals: Advances in Chemistry*. 1:244-49.

[256] Zahm, S. H., Weisenburger, D. D., et al. (1990). "A case-control study of non-Hodgkin's lymphoma and the herbicide 2,4-dichlorophenoxyacetic acid (2, 4-d) in eastern Nebraska." *Epidemiology*. 1(5):349-56.

[257] Burns, C. J., Beard, K. K., Cartmill, J. B. (2001). "Mortality in chemical workers potentially exposed to 2,4-dichlorophenoxyacetic acid (2,4-D) 1945-94: an update." *Occup Environ Med*. 58(1):24-30.

[258] Burns, supra.

[259] Lerda, D., Rizzi, R. (1991). "Study of reproductive function in persons occupationally exposed to 2,4-dichlorophenoxyacetic acid (2,4-D)." *Muta Res*. 262(1):47-50.

[260] Duhigg, Charles. "Debating How Much Weed Killer Is Safe in Your Water Glass." *New York Times*, August 22, 2009.

[261] Ackerman, F. (2007). "The economics of atrazine." *Internat J Occup Environ Health*. 13(4):437-45.

[262] Walsh, Edward. "EPA Stops Short of Banning Herbicide." *Washington Post*, February 1, 2003.

[263] Reregistration Eligibility Decision (RED): Glyphosate; EPA-738-R-93-014; U.S. Environmental Protection Agency, Office of Prevention, Pesticides, and Toxic Substances, Office of Pesticide Programs, U.S. Government Printing Office: Washington, DC, 1993.

[264] Richard, S., Moslemi, S., et al. (2005). "Differential effects of glyphosate and Roundup on human placental cells and aromatase." *Environ Health Perspect*. 113(6):716-20.

[265] Brunel University. (August 7, 2008). "Use of Cleaning Products During Pregnancy Increases Risk of Asthma in Young Children." *ScienceDaily*. Retrieved October 31, 2013 from http://www.sciencedaily.com / releases/2008/08/080806154716.htm.

[266] Oregon Department of Environmental Quality. Retrieved November 1, 2013 from http://www.deq.state.or.us/programs/sustainability/toxicalternatives.htm.

[267] Grassroots Environmental Education. "Questions and Answers About Green Cleaning." *The ChildSafe School Program*. 2010.

[268] Landrigan, P. J., Garg, A. (2002). "Chronic effects of toxic environmental exposures on children's health." *J Toxicol Clin Toxicol*. 40(4):449-56.

[269] Grandjean, Phillippe. *Only One Chance: How Environmental Pollution Impairs Brain Development—and How to Protect the Brains of the Next Generation*. New York: Oxford University Press, 2013: 10.

[270] International Chewing Gum Association FAQ. Retrieved November 1, 2013 from http://www.gumassociation.org/index.cfm/facts-figures/frequently-asked-questions/how-can-i-best-get-gum-unstuck-if-it-ends-up-where-it-doesnt-belong1.

[271] Schachner, Lawrence A., and Ronald C. Hansen, eds. *Pediatric Dermatology*. Mosby, 2011: 121.

[272] Wynters, Sharon and Burton Goldberg. *The Pure Cure: A Complete Guide to Freeing Your Life from Dangerous Toxins*. Berkeley, CA: Soft Skull Press, 2012: 211-12. Bauer, *Diaper Free: The Gentle Wisdom of Natural Infant Hygiene*. New York: Plume, 2006: Chapter 1.

[273] Clawson, Julie. *Everyday Justice: The Global Impact of Our Daily Choices*. Downer's Grove, IL: InterVarsity Press, 2009: 153.

[274] Ashton, Karen and Green, Elizabeth Salter. *The Toxic Consumer: Living Healthy in a Hazardous World*. New York: Sterling, 2008: 73-74.

[275] The average was 3,796 diapers, to be exact. An updated lifecycle assessment study for disposable and reusable nappies. Department for Environment Food and Rural Affairs at 16. Environment Agency, 2008. Available at http://randd.defra.gov.uk/Document.aspx?Document=WR0705_7589_FRP.pdf.

[276] 3.6 million pounds of disposable diapers were dumped in our landfills in 2011, but that number includes diapers marketed for adult incontinence. Municipal Solid Waste in the United States: 2011 Facts and Figures at 75. U.S. Environmental Protection Agency, 2013. http://www.epa.gov/epawaste/nonhaz/municipal/pubs/MSWcharacterization_fnl_060713_2_rpt.pdf.

[277] Sathyanarayana, S., Karr, C. J., et al. (February 2008). "Baby care products: possible sources of infant phthalate exposure." *Pediatrics*. 121(2):e260-68.

[278] Lumsden, Hilary and Debbie Holmes, eds. *Care of the Newborn by Ten Teachers*. London: Taylor & Francis Group, 2010: 83-84.

[279] Greene, Alan. *Raising Baby Green: The Earth-Friendly Guide to Pregnancy, Childbirth, and Baby Care*. San Francisco: Jossey-Bass, 2007: 172.

[280] Wynters and Goldberg, supra at 177.

[281] American Forest and Paper Association (Garner, J. W. Energy Conservation Practices Offer Environmental and Cost Benefits. Pulp & Paper, October 2002).

[282] Kerr, Bob and Mary Lee Hall. *1991–1992 Green Index: A State-By-State Guide to the Nation's Environmental Health at 6*. Washington DC: Island Press, 1991.

[283] U.S. Congress, Office of Technology Assessment, Technologies for Dioxin in the Manufacture of Bleached Wood Pulp, OTA-BP-O-54 at 5 (Washington, DC: U.S. Government Printing Office, May 1989).

Chelliah, R. S., Appasamy, P. P., et al. (2007). "Ecotaxes: On polluting inputs and outputs." *Academic Foundation*. 125-26.

[284] Mocarelli, P., Brambilla, P., et al. (August 1996). "Change in sex ratio with exposure to dioxin." *Lancet*. 348(9024):409.

[285] Mandal, P. K. (2005). "Dioxin: a review of its environmental effects and its aryl hydrocarbon receptor biology." *J Comp Physiol B*. 175:221-230.

Ngo, A. D., et al. "Association between Agent Orange and birth defects: systematic review and meta-analysis." *Int J Epidemiol*. 35(5):1220-30.

Phillips, Robert, and Glenda Motta. *Coping with Endometriosis: A Practical Guide*. New York: Avery, 2000: 10.

[286] Park, Chris. *The Environment: Principles and Applications*. 2nd ed. New York: Routledge, 2001: 96.

[287] Phillips, supra.

[288] Park, supra at 120.

[289] Park, supra at 256.

[290] Matsushima, A., Kakuta, Y., et al. (2007). "Structural evidence for endocrine disruptor bisphenol A binding to human nuclear receptor ERR gamma." *J Biochem.* 142(4):517-24.

[291] Rubin, B. S. (2011). "Bisphenol A: an endocrine disruptor with widespread exposure and multiple effects." *J Steroid Biochem Mol Bio.* 127:27-34.

[292] Okada, H., Tokunaga, T., et al. (2008). "Direct evidence revealing structural elements essential for the high binding ability of bisphenol A to human estrogen-related receptor-gamma." *Environ Health Perspect.* 116(1):32-38.

[293] *Kirk-Othmer Encyclopedia of Chemical Technology.* 5th ed. 5(8). Hoboken, NJ: John Wiley & Sons, 2006.

[294] Garnder, Amanda. "Studies Report More Harmful Effects From BPA." *U.S. News & World Report.* June 10, 2009.

[295] Fukazawa, H., Hoshino, K., et al. (2001). "Identification and quantification of chlorinated bisphenol A in wastewater from wastepaper recycling plants." *Chemosphere.* 44(5):973-79.

[296] vom Saal F. S., Akingbemi, B. T., et al. (2007). "Chapel Hill bisphenol A expert panel consensus statement: integration of mechanisms, effects in animals and potential to impact human health at current levels of exposure." *Reprod Toxicol.* 24(2):131-38.

[297] Vogel, S. (2009). "The politics of plastics: the making and unmaking of bisphenol A 'safety.'" *Am J Public Health.* 99(S3):559-66.

[298] Sax, L. (April 2010). "Polyethylene terephthalate may yield endocrine disruptors." *Environ Health Perspect.* 118(4):445-48.

[299] Kuruto-Niwa, R., Nozawa, R., et al. (2005). "Estrogenic activity of alkylphenols, bisphenol S, and their chlorinated derivatives using a GFP expression system." *Environ Toxicol Pharmacol.* 19(1):121-30.

[300] Viñas, R., Watson, C. S. (2013). "Bisphenol S disrupts estradiol-induced nongenomic signaling in a rat pituitary cell line: effects on cell functions." *Environ Health Perspect.* 121(3):253-58.

[301] Danzl, E., Sei, K., et al. (April 2009). "Biodegradation of bisphenol A, bisphenol F and bisphenol S in seawater." *Int J Environ Res Public Health*. 6(4):1472-84.

[302] Jakszyn, P., Gonzalez, C. A. (2006). "Nitrosamine and related food intake and gastric and oesophageal cancer risk: a systematic review of the epidemiological evidence." *World J Gastroenterol*. 12(27):4296-303.

[303] "Municipal Solid Waste Generation, Recycling, and Disposal in the United States: Facts and Figures for 2011." U.S. Environmental Protection Agency.

[304] Petersen J. H., Jensen L. K. (November 2010). "Phthalates and food-contact materials: enforcing the 2008 European Union plastics legislation." *Food Addit Contam Part A Chem Anal Control Expo Risk Assess*. 27(11):1608-16.

[305] The Toilet Paper Encyclopedia. http://encyclopedia.toiletpaperworld.com/toilet-paper-facts.

[306] Rogers, Elizabeth, and Kostigen, Thomas M. *The Green Book: The Everyday Guide to Saving the Planet One Simple Step at a Time*. New York: Three Rivers Press, 2007: 57.

[307] Rogers, supra.

[308] Rogers, supra.

[309] Rogers, supra.

[310] Winter, Ruth. *A Consumer's Dictionary of Food Additives: Descriptions in Plain English of More than 12,000 Ingredients Both Harmful and Desirable Found in Foods*. 7th Edition. New York: Three Rivers Press, 2009: 109.

[311] Tugend, Alena. "If Your Appliances Are Avocado, They're Probably Not Green." *New York Times*, May 10, 2008.

[312] Consumer Expenditures 2011. BLS Reports April 2013. U.S. Bureau of Labor Statistics. http://www.bls.gov/cex/csxann11.pdf.

INDEX

ACKNOWLEDGMENTS

Homemade Cleaners: A Recipe for an Impassioned Collaboration

1 PART LOVE AND SUPPORT FROM OUR FAMILIES: You inspire us to leave this Earth better than we found it.

Dionna is grateful to Tom for making her happiness his goal (how lucky she is!); to Kieran and Ailia, without whom the sun would have nowhere to rise and set; to Tammy, for "which" hunts and giggle fests; to Shawna, for getting it; and to Mom and Dad, for believing in her, always.

Mandy would like to thank her husband and best friend, Bart, who has supported her throughout the years in her many endeavors and has been willing to listen to research and change preconceived societal views. Thanks also to Bart for his help in formatting. She is thankful for her children, Eoin, Eilis, Eamon, and Ellery, who have allowed her to grow in so many ways she never imagined and continue to challenge her to explore the world around her. They are the smile in her day and the hug in her heart.

1 PART CONSTRUCTIVE CRITICISM AND KNOW-HOW FROM A SAVVY EDITORIAL AND MARKETING TEAM: Your insight gave us the clarity we needed to make this book a reality. We are grateful to Alice

Riegert, Katherine Furman, Kathy Kaiser, Kourtney Joy, and all of the staff at Ulysses Press who helped with *Homemade Cleaners*.

1 PART INSPIRATION FROM OUR FRIENDS AND READERS: We laugh with you, we learn from you, we write for you. We'd like to mention in particular our friends who comprise the community of volunteers from Natural Parents Network (www.NaturalParentsNetwork.com), our friends in Kansas City, and the readers of CodeNameMama.com, LivingPeacefullyWithChildren.com, and Natural Parents Network and their respective communities on Facebook.

1 PART GRATITUDE TO THOSE WHO HAVE GONE BEFORE US: Thank you to the scientists who published the studies we read in research for *Homemade Cleaners* and to the authors and bloggers who have written about and normalized "green" cleaning. Thanks also goes to the generations past whose experience and wisdom about simple, effective cleaning has been passed down. We haven't invented something new, we have just embraced the previous knowledge of others, now backed up with scientific studies.

1 PART FRIENDSHIP: between two mothers who love learning and creating and who are striving to make a better place for their children and others.

ADD IN SOME DETECTIVE WORK AND RESEARCH SKILLS, AND THE RESULT IS *HOMEMADE CLEANERS*: our labor of love that can help you make simple changes toward a less toxic home.

ABOUT THE AUTHORS

MANDY O'BRIEN is a biologist turned stay-at-home mom with an additional degree in environmental studies and graduate studies in ecology, among other subjects. An advocate of human rights and consensual living, she is working toward a more sustainable, simple-living lifestyle. Mandy is married to her college sweetheart and is mother to four fantastic children. She advocates for consensual and natural living at her website, *Living Peacefully with Children* (www.livingpeacefullywithchildren.com), and volunteers for *Natural Parents Network* (www.NaturalParentsNetwork.com), while organizing parenting and homeschooling events and living a mindful life with her family. The O'Briens live in Wisconsin.

DIONNA FORD is a lawyer turned work-at-home mama. Her advocacy for natural parenting branched quite naturally into an interest in sustainable living. Dionna shares her passion for attachment parenting and natural family living at her website, *Code Name: Mama* (www.codenamemama.com). She cofounded two other websites for parents, *Natural Parents Network* and *Nursing Freedom* (www.nursingfreedom.org). She makes her home in Kansas City, where she is active in the parenting and homeschooling communities.